第2版（下册）

流利的Python语言（影印版）

Fluent Python

Luciano Ramalho 著

Beijing · Boston · Farnham · Sebastopol · Tokyo

O'Reilly Media, Inc.授权东南大学出版社出版

南京 东南大学出版社

图书在版编目(CIP)数据

流利的 Python 语言：第 2 版 = Fluent Python, 2E：
英文 /（巴西）卢西亚诺·拉马略著. —影印本. —南
京：东南大学出版社,2022.9
ISBN 978 - 7 - 5766 - 0147 - 3

Ⅰ.①流… Ⅱ.①卢… Ⅲ.①软件工具−程序设计−
英文 Ⅳ.①TP311.561

中国版本图书馆 CIP 数据核字(2022)第 110841 号
图字：10 - 2020 - 387 号

流利的 Python 语言 第 2 版（影印版）

著　　者：Luciano Ramalho
责任编辑：张　烨　封面设计：Karen Montgomery,张　健　　责任印制：周荣虎
出版发行：东南大学出版社
社　　址：南京四牌楼 2 号　　邮编：210096　　电话：025-83793330
网　　址：http://www.seupress.com
电子邮件：press@ seupress.com
经　　销：全国各地新华书店
印　　刷：常州市武进第三印刷有限公司
开　　本：787mm×980mm　1/16
印　　张：64.5
字　　数：1263 千
版　　次：2022 年 9 月第 1 版
版　　次：2022 年 9 月第 1 次印刷
书　　号：ISBN 978 - 7 - 5766 - 0147 - 3
定　　价：198.00 元(上下册)

本社图书若有印装质量问题,请直接与营销部联系。电话(传真)：025 - 83791830

Para Marta, com todo o meu amor.

Table of Contents

Part III.   Classes and Protocols

Part V.   Metaprogramming

## PART IV

Control Flow

## CHAPTER 17

Iterators, Generators, and Classic Coroutines

When I see patterns in my programs, I consider it a sign of trouble. The shape of a program should reflect only the problem it needs to solve. Any other regularity in the code is a sign, to me at least, that I'm using abstractions that aren't powerful enough—often that I'm generating by hand the expansions of some macro that I need to write.

—Paul Graham, Lisp hacker and venture capitalist[1]

Iteration is fundamental to data processing: programs apply computations to data series, from pixels to nucleotides. If the data doesn't fit in memory, we need to fetch the items *lazily*—one at a time and on demand. That's what an iterator does. This chapter shows how the *Iterator* design pattern is built into the Python language so you never need to code it by hand.

Every standard collection in Python is *iterable*. An *iterable* is an object that provides an *iterator*, which Python uses to support operations like:

- for loops
- List, dict, and set comprehensions
- Unpacking assignments
- Construction of collection instances

---

1 From "Revenge of the Nerds" (*https://fpy.li/17-1*), a blog post.

This chapter covers the following topics:

- How Python uses the `iter()` built-in function to handle iterable objects
- How to implement the classic Iterator pattern in Python
- How the classic Iterator pattern can be replaced by a generator function or generator expression
- How a generator function works in detail, with line-by-line descriptions
- Leveraging the general-purpose generator functions in the standard library
- Using `yield from` expressions to combine generators
- Why generators and classic coroutines look alike but are used in very different ways and should not be mixed

What's New in This Chapter

"Subgenerators with yield from" on page 636 grew from one to six pages. It now includes simpler experiments demonstrating the behavior of generators with `yield from`, and an example of traversing a tree data structure, developed step-by-step.

New sections explain the type hints for `Iterable`, `Iterator`, and `Generator` types.

The last major section of this chapter, "Classic Coroutines" on page 645, is a 9-page introduction to a topic that filled a 40-page chapter in the first edition. I updated and moved the "Classic Coroutines" chapter to a post in the companion website (*https://fpy.li/oldcoro*) because it was the most challenging chapter for readers, but its subject matter is less relevant after Python 3.5 introduced native coroutines—which we'll study in Chapter 21.

We'll get started studying how the `iter()` built-in function makes sequences iterable.

A Sequence of Words

We'll start our exploration of iterables by implementing a `Sentence` class: you give its constructor a string with some text, and then you can iterate word by word. The first version will implement the sequence protocol, and it's iterable because all sequences are iterable—as we've seen since Chapter 1. Now we'll see exactly why.

Example 17-1 shows a `Sentence` class that extracts words from a text by index.

Example 17-1. sentence.py: a Sentence as a sequence of words

```
import re
import reprlib
```

```
RE_WORD = re.compile(r'\w+')


class Sentence:

    def __init__(self, text):
        self.text = text
        self.words = RE_WORD.findall(text)  ❶

    def __getitem__(self, index):
        return self.words[index]  ❷

    def __len__(self):  ❸
        return len(self.words)

    def __repr__(self):
        return 'Sentence(%s)' % reprlib.repr(self.text)  ❹
```

❶ .findall returns a list with all nonoverlapping matches of the regular expression, as a list of strings.

❷ self.words holds the result of .findall, so we simply return the word at the given index.

❸ To complete the sequence protocol, we implement __len__ although it is not needed to make an iterable.

❹ reprlib.repr is a utility function to generate abbreviated string representations of data structures that can be very large.[2]

By default, reprlib.repr limits the generated string to 30 characters. See the console session in Example 17-2 to see how Sentence is used.

Example 17-2. Testing iteration on a Sentence instance

```
>>> s = Sentence('"The time has come," the Walrus said,')  ❶
>>> s
Sentence('"The time ha... Walrus said,')  ❷
>>> for word in s:  ❸
...     print(word)
The
time
has
come
```

---

2 We first used reprlib in "Vector Take #1: Vector2d Compatible" on page 401.

```
the
Walrus
said
>>> list(s)  ❹
['The', 'time', 'has', 'come', 'the', 'Walrus', 'said']
```

❶   A sentence is created from a string.

❷   Note the output of __repr__ using . . . generated by reprlib.repr.

❸   Sentence instances are iterable; we'll see why in a moment.

❹   Being iterable, Sentence objects can be used as input to build lists and other iterable types.

In the following pages, we'll develop other Sentence classes that pass the tests in Example 17-2. However, the implementation in Example 17-1 is different from the others because it's also a sequence, so you can get words by index:

```
>>> s[0]
'The'
>>> s[5]
'Walrus'
>>> s[-1]
'said'
```

Python programmers know that sequences are iterable. Now we'll see precisely why.

Why Sequences Are Iterable: The iter Function

Whenever Python needs to iterate over an object x, it automatically calls iter(x).

The iter built-in function:

1. Checks whether the object implements __iter__, and calls that to obtain an iterator.
2. If __iter__ is not implemented, but __getitem__ is, then iter() creates an iterator that tries to fetch items by index, starting from 0 (zero).
3. If that fails, Python raises TypeError, usually saying 'C' object is not iterable, where C is the class of the target object.

That is why all Python sequences are iterable: by definition, they all implement __getitem__. In fact, the standard sequences also implement __iter__, and yours should too, because iteration via __getitem__ exists for backward compatibility and may be gone in the future—although it is not deprecated as of Python 3.10, and I doubt it will ever be removed.

As mentioned in "Python Digs Sequences" on page 438, this is an extreme form of duck typing: an object is considered iterable not only when it implements the special method __iter__, but also when it implements __getitem__. Take a look:

```
>>> class Spam:
...     def __getitem__(self, i):
...         print('->', i)
...         raise IndexError()
...
>>> spam_can = Spam()
>>> iter(spam_can)
<iterator object at 0x10a878f70>
>>> list(spam_can)
-> 0
[]
>>> from collections import abc
>>> isinstance(spam_can, abc.Iterable)
False
```

If a class provides __getitem__, the iter() built-in accepts an instance of that class as iterable and builds an iterator from the instance. Python's iteration machinery will call __getitem__ with indexes starting from 0, and will take an IndexError as a signal that there are no more items.

Note that although spam_can is iterable (its __getitem__ could provide items), it is not recognized as such by an isinstance against abc.Iterable.

In the goose-typing approach, the definition for an iterable is simpler but not as flexible: an object is considered iterable if it implements the __iter__ method. No subclassing or registration is required, because abc.Iterable implements the __subclasshook__, as seen in "Structural Typing with ABCs" on page 466. Here is a demonstration:

```
>>> class GooseSpam:
...     def __iter__(self):
...         pass
...
>>> from collections import abc
>>> issubclass(GooseSpam, abc.Iterable)
True
>>> goose_spam_can = GooseSpam()
>>> isinstance(goose_spam_can, abc.Iterable)
True
```

As of Python 3.10, the most accurate way to check whether an object x is iterable is to call iter(x) and handle a TypeError exception if it isn't. This is more accurate than using isinstance(x, abc.Iterable), because iter(x) also considers the legacy __getitem__ method, while the Iterable ABC does not.

Explicitly checking whether an object is iterable may not be worthwhile if right after the check you are going to iterate over the object. After all, when the iteration is attempted on a noniterable, the exception Python raises is clear enough: `TypeError: 'C' object is not iterable`. If you can do better than just raising `TypeError`, then do so in a `try/except` block instead of doing an explicit check. The explicit check may make sense if you are holding on to the object to iterate over it later; in this case, catching the error early makes debugging easier.

The `iter()` built-in is more often used by Python itself than by our own code. There's a second way we can use it, but it's not widely known.

Using iter with a Callable

We can call `iter()` with two arguments to create an iterator from a function or any callable object. In this usage, the first argument must be a callable to be invoked repeatedly (with no arguments) to produce values, and the second argument is a *sentinel* (*https://fpy.li/17-2*): a marker value which, when returned by the callable, causes the iterator to raise `StopIteration` instead of yielding the sentinel.

The following example shows how to use `iter` to roll a six-sided die until a 1 is rolled:

```
>>> def d6():
...     return randint(1, 6)
...
>>> d6_iter = iter(d6, 1)
>>> d6_iter
<callable_iterator object at 0x10a245270>
>>> for roll in d6_iter:
...     print(roll)
...
4
3
6
3
```

Note that the `iter` function here returns a `callable_iterator`. The `for` loop in the example may run for a very long time, but it will never display 1, because that is the sentinel value. As usual with iterators, the `d6_iter` object in the example becomes useless once exhausted. To start over, we must rebuild the iterator by invoking `iter()` again.

The documentation for `iter` (*https://fpy.li/17-3*) includes the following explanation and example code:

> One useful application of the second form of `iter()` is to build a block-reader. For example, reading fixed-width blocks from a binary database file until the end of file is reached:

```
from functools import partial

with open('mydata.db', 'rb') as f:
    read64 = partial(f.read, 64)
    for block in iter(read64, b''):
        process_block(block)
```

For clarity, I've added the `read64` assignment, which is not in the original example (*https://fpy.li/17-3*). The `partial()` function is necessary because the callable given to `iter()` must not require arguments. In the example, an empty bytes object is the sentinel, because that's what `f.read` returns when there are no more bytes to read.

The next section details the relationship between iterables and iterators.

Iterables Versus Iterators

From the explanation in "Why Sequences Are Iterable: The iter Function" on page 600 we can extrapolate a definition:

iterable

> Any object from which the `iter` built-in function can obtain an iterator. Objects implementing an `__iter__` method returning an *iterator* are iterable. Sequences are always iterable, as are objects implementing a `__getitem__` method that accepts 0-based indexes.

It's important to be clear about the relationship between iterables and iterators: Python obtains iterators from iterables.

Here is a simple `for` loop iterating over a `str`. The `str` `'ABC'` is the iterable here. You don't see it, but there is an iterator behind the curtain:

```
>>> s = 'ABC'
>>> for char in s:
...     print(char)
...
A
B
C
```

If there was no `for` statement and we had to emulate the `for` machinery by hand with a `while` loop, this is what we'd have to write:

```
>>> s = 'ABC'
>>> it = iter(s)   ❶
>>> while True:
...     try:
...         print(next(it))   ❷
...     except StopIteration:   ❸
...         del it   ❹
...         break   ❺
```

```
...
A
B
C
```

❶ Build an iterator `it` from the iterable.

❷ Repeatedly call `next` on the iterator to obtain the next item.

❸ The iterator raises `StopIteration` when there are no further items.

❹ Release reference to `it`—the iterator object is discarded.

❺ Exit the loop.

`StopIteration` signals that the iterator is exhausted. This exception is handled internally by the `iter()` built-in that is part of the logic of `for` loops and other iteration contexts like list comprehensions, iterable unpacking, etc.

Python's standard interface for an iterator has two methods:

`__next__`
> Returns the next item in the series, raising `StopIteration` if there are no more.

`__iter__`
> Returns `self`; this allows iterators to be used where an iterable is expected, for example, in a `for` loop.

That interface is formalized in the `collections.abc.Iterator` ABC, which declares the `__next__` abstract method, and subclasses `Iterable`—where the abstract `__iter__` method is declared. See Figure 17-1.

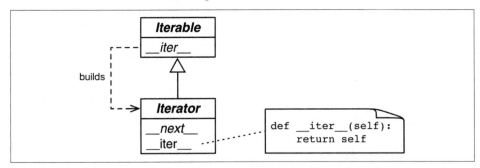

Figure 17-1. The Iterable and Iterator ABCs. Methods in italic are abstract. A concrete Iterable.__iter__ should return a new Iterator instance. A concrete Iterator must implement __next__. The Iterator.__iter__ method just returns the instance itself.

The source code for collections.abc.Iterator is in Example 17-3.

Example 17-3. abc.Iterator class; extracted from Lib/_collections_abc.py (https://fpy.li/17-5)

```
class Iterator(Iterable):

    __slots__ = ()

    @abstractmethod
    def __next__(self):
        'Return the next item from the iterator. When exhausted, raise StopIteration'
        raise StopIteration

    def __iter__(self):
        return self

    @classmethod
    def __subclasshook__(cls, C):     ❶
        if cls is Iterator:
            return _check_methods(C, '__iter__', '__next__')     ❷
        return NotImplemented
```

❶ __subclasshook__ supports structural type checks with isinstance and issub class. We saw it in "Structural Typing with ABCs" on page 466.

❷ _check_methods traverses the __mro__ of the class to check whether the methods are implemented in its base classes. It's defined in that same *Lib/_collections_abc.py* module. If the methods are implemented, the C class will be recognized as a virtual subclass of Iterator. In other words, issubclass(C, Iterable) will return True.

> The Iterator ABC abstract method is it.__next__() in Python 3 and it.next() in Python 2. As usual, you should avoid calling special methods directly. Just use the next(it): this built-in function does the right thing in Python 2 and 3—which is useful for those migrating codebases from 2 to 3.

The *Lib/types.py* (*https://fpy.li/17-6*) module source code in Python 3.9 has a comment that says:

```
# Iterators in Python aren't a matter of type but of protocol. A large
# and changing number of builtin types implement *some* flavor of
# iterator. Don't check the type! Use hasattr to check for both
# "__iter__" and "__next__" attributes instead.
```

In fact, that's exactly what the __subclasshook__ method of the abc.Iterator ABC does.

 Given the advice from *Lib/types.py* and the logic implemented in *Lib/_collections_abc.py*, the best way to check if an object x is an iterator is to call isinstance(x, abc.Iterator). Thanks to Iterator.__subclasshook__, this test works even if the class of x is not a real or virtual subclass of Iterator.

Back to our Sentence class from Example 17-1, you can clearly see how the iterator is built by iter() and consumed by next() using the Python console:

```
>>> s3 = Sentence('Life of Brian')  ❶
>>> it = iter(s3)  ❷
>>> it  # doctest: +ELLIPSIS
<iterator object at 0x...>
>>> next(it)  ❸
'Life'
>>> next(it)
'of'
>>> next(it)
'Brian'
>>> next(it)  ❹
Traceback (most recent call last):
  ...
StopIteration
>>> list(it)  ❺
[]
>>> list(iter(s3))  ❻
['Life', 'of', 'Brian']
```

❶ Create a sentence s3 with three words.

❷ Obtain an iterator from s3.

❸ next(it) fetches the next word.

❹ There are no more words, so the iterator raises a StopIteration exception.

❺ Once exhausted, an iterator will always raise StopIteration, which makes it look like it's empty.

❻ To go over the sentence again, a new iterator must be built.

Because the only methods required of an iterator are __next__ and __iter__, there is no way to check whether there are remaining items, other than to call next() and catch StopIteration. Also, it's not possible to "reset" an iterator. If you need to start over, you need to call `iter()` on the iterable that built the iterator in the first place. Calling `iter()` on the iterator itself won't help either, because—as mentioned—`Iterator.__iter__` is implemented by returning `self`, so this will not reset a depleted iterator.

That minimal interface is sensible, because in reality not all iterators are resettable. For example, if an iterator is reading packets from the network, there's no way to rewind it.[3]

The first version of `Sentence` from Example 17-1 was iterable thanks to the special treatment the `iter()` built-in gives to sequences. Next, we will implement `Sentence` variations that implement `__iter__` to return iterators.

Sentence Classes with __iter__

The next variations of `Sentence` implement the standard iterable protocol, first by implementing the Iterator design pattern, and then with generator functions.

Sentence Take #2: A Classic Iterator

The next `Sentence` implementation follows the blueprint of the classic Iterator design pattern from the *Design Patterns* book. Note that it is not idiomatic Python, as the next refactorings will make very clear. But it is useful to show the distinction between an iterable collection and an iterator that works with it.

The `Sentence` class in Example 17-4 is iterable because it implements the `__iter__` special method, which builds and returns a `SentenceIterator`. That's how an iterable and an iterator are related.

Example 17-4. sentence_iter.py: `Sentence` implemented using the Iterator pattern

```
import re
import reprlib

RE_WORD = re.compile(r'\w+')


class Sentence:

    def __init__(self, text):
        self.text = text
        self.words = RE_WORD.findall(text)
```

---

3 Thanks to tech reviewer Leonardo Rochael for this fine example.

```
    def __repr__(self):
        return f'Sentence({reprlib.repr(self.text)})'

    def __iter__(self):  ❶
        return SentenceIterator(self.words)  ❷


class SentenceIterator:

    def __init__(self, words):
        self.words = words  ❸
        self.index = 0  ❹

    def __next__(self):
        try:
            word = self.words[self.index]  ❺
        except IndexError:
            raise StopIteration()  ❻
        self.index += 1  ❼
        return word  ❽

    def __iter__(self):  ❾
        return self
```

❶ The __iter__ method is the only addition to the previous Sentence implementation. This version has no __getitem__, to make it clear that the class is iterable because it implements __iter__.

❷ __iter__ fulfills the iterable protocol by instantiating and returning an iterator.

❸ SentenceIterator holds a reference to the list of words.

❹ self.index determines the next word to fetch.

❺ Get the word at self.index.

❻ If there is no word at self.index, raise StopIteration.

❼ Increment self.index.

❽ Return the word.

❾ Implement self.__iter__.

The code in Example 17-4 passes the tests in Example 17-2.

Note that implementing __iter__ in SentenceIterator is not actually needed for this example to work, but it is the right thing to do: iterators are supposed to implement both __next__ and __iter__, and doing so makes our iterator pass the issubclass(SentenceIterator, abc.Iterator) test. If we had subclassed SentenceIterator from abc.Iterator, we'd inherit the concrete abc.Itera tor.__iter__ method.

That is a lot of work (for us spoiled Python programmers, anyway). Note how most code in SentenceIterator deals with managing the internal state of the iterator. Soon we'll see how to avoid that bookkeeping. But first, a brief detour to address an implementation shortcut that may be tempting, but is just wrong.

Don't Make the Iterable an Iterator for Itself

A common cause of errors in building iterables and iterators is to confuse the two. To be clear: iterables have an __iter__ method that instantiates a new iterator every time. Iterators implement a __next__ method that returns individual items, and an __iter__ method that returns self.

Therefore, iterators are also iterable, but iterables are not iterators.

It may be tempting to implement __next__ in addition to __iter__ in the Sentence class, making each Sentence instance at the same time an iterable and iterator over itself. But this is rarely a good idea. It's also a common antipattern, according to Alex Martelli who has a lot of experience reviewing Python code at Google.

The "Applicability" section about the Iterator design pattern in the *Design Patterns* book says:

> Use the Iterator pattern
>
> - to access an aggregate object's contents without exposing its internal representation.
> - to support multiple traversals of aggregate objects.
> - to provide a uniform interface for traversing different aggregate structures (that is, to support polymorphic iteration).

To "support multiple traversals," it must be possible to obtain multiple independent iterators from the same iterable instance, and each iterator must keep its own internal state, so a proper implementation of the pattern requires each call to iter(my_itera ble) to create a new, independent, iterator. That is why we need the SentenceItera tor class in this example.

Now that the classic Iterator pattern is properly demonstrated, we can let it go. Python incorporated the yield keyword from Barbara Liskov's CLU language (*https://fpy.li/17-7*), so we don't need to "generate by hand" the code to implement iterators.

The next sections present more idiomatic versions of Sentence.

Sentence Take #3: A Generator Function

A Pythonic implementation of the same functionality uses a generator, avoiding all the work to implement the SentenceIterator class. A proper explanation of the generator comes right after Example 17-5.

Example 17-5. sentence_gen.py: Sentence implemented using a generator

```
import re
import reprlib

RE_WORD = re.compile(r'\w+')


class Sentence:

    def __init__(self, text):
        self.text = text
        self.words = RE_WORD.findall(text)

    def __repr__(self):
        return 'Sentence(%s)' % reprlib.repr(self.text)

    def __iter__(self):
        for word in self.words:      ❶
            yield word      ❷
        ❸

# done!  ❹
```

❶ Iterate over self.words.

❷ Yield the current word.

❸ Explicit return is not necessary; the function can just "fall through" and return automatically. Either way, a generator function doesn't raise StopIteration: it simply exits when it's done producing values.[4]

---

4 When reviewing this code, Alex Martelli suggested the body of this method could simply be return iter(self.words). He is right: the result of calling self.words.__iter__() would also be an iterator, as it should be. However, I used a for loop with yield here to introduce the syntax of a generator function, which requires the yield keyword, as we'll see in the next section. During review of the second edition of this book, Leonardo Rochael suggested yet another shortcut for the body of __iter__: yield from self.words. We'll also cover yield from later in this chapter.

---

❹ No need for a separate iterator class!

Here again we have a different implementation of Sentence that passes the tests in Example 17-2.

Back in the Sentence code in Example 17-4, __iter__ called the SentenceIterator constructor to build an iterator and return it. Now the iterator in Example 17-5 is in fact a generator object, built automatically when the __iter__ method is called, because __iter__ here is a generator function.

A full explanation of generators follows.

How a Generator Works

Any Python function that has the yield keyword in its body is a generator function: a function which, when called, returns a generator object. In other words, a generator function is a generator factory.

The only syntax distinguishing a plain function from a generator function is the fact that the latter has a yield keyword somewhere in its body. Some argued that a new keyword like gen should be used instead of def to declare generator functions, but Guido did not agree. His arguments are in PEP 255 — Simple Generators (*https://fpy.li/pep255*).[5]

Example 17-6 shows the behavior of a simple generator function.[6]

Example 17-6. A generator function that yields three numbers

```
>>> def gen_123():
...     yield 1   ❶
...     yield 2
...     yield 3
...
>>> gen_123  # doctest: +ELLIPSIS
<function gen_123 at 0x...>   ❷
>>> gen_123()   # doctest: +ELLIPSIS
<generator object gen_123 at 0x...>   ❸
>>> for i in gen_123():   ❹
...     print(i)
```

---

5 Sometimes I add a gen prefix or suffix when naming generator functions, but this is not a common practice. And you can't do that if you're implementing an iterable, of course: the necessary special method must be named __iter__.

6 Thanks to David Kwast for suggesting this example.

```
1
2
3
>>> g = gen_123()  ❺
>>> next(g)  ❻
1
>>> next(g)
2
>>> next(g)
3
>>> next(g)  ❼
Traceback (most recent call last):
  ...
StopIteration
```

❶ The body of a generator function often has yield inside a loop, but not necessarily; here I just repeat yield three times.

❷ Looking closely, we see gen_123 is a function object.

❸ But when invoked, gen_123() returns a generator object.

❹ Generator objects implement the Iterator interface, so they are also iterable.

❺ We assign this new generator object to g, so we can experiment with it.

❻ Because g is an iterator, calling next(g) fetches the next item produced by yield.

❼ When the generator function returns, the generator object raises StopIteration.

A generator function builds a generator object that wraps the body of the function. When we invoke next() on the generator object, execution advances to the next yield in the function body, and the next() call evaluates to the value yielded when the function body is suspended. Finally, the enclosing generator object created by Python raises StopIteration when the function body returns, in accordance with the Iterator protocol.

I find it helpful to be rigorous when talking about values obtained from a generator. It's confusing to say a generator "returns" values. Functions return values. Calling a generator function returns a generator. A generator yields values. A generator doesn't "return" values in the usual way: the return statement in the body of a generator function causes StopIteration to be raised by the generator object. If you return x in the generator, the caller can retrieve the value of x from the StopIteration exception, but usually that is done automatically using the yield from syntax, as we'll see in "Returning a Value from a Coroutine" on page 650.

Example 17-7 makes the interaction between a for loop and the body of the function more explicit.

Example 17-7. A generator function that prints messages when it runs

```
>>> def gen_AB():
...     print('start')
...     yield 'A'           ❶
...     print('continue')
...     yield 'B'           ❷
...     print('end.')       ❸
...
>>> for c in gen_AB():     ❹
...     print('-->', c)    ❺
...
start       ❻
--> A       ❼
continue    ❽
--> B       ❾
end.        ❿
>>>         ⓫
```

❶ The first implicit call to next() in the for loop at ❹ will print 'start' and stop at the first yield, producing the value 'A'.

❷ The second implicit call to next() in the for loop will print 'continue' and stop at the second yield, producing the value 'B'.

❸ The third call to next() will print 'end.' and fall through the end of the function body, causing the generator object to raise StopIteration.

❹ To iterate, the for machinery does the equivalent of g = iter(gen_AB()) to get a generator object, and then next(g) at each iteration.

❺ The loop prints `-->` and the value returned by `next(g)`. This output will appear only after the output of the `print` calls inside the generator function.

❻ The text `start` comes from `print('start')` in the generator body.

❼ `yield 'A'` in the generator body yields the value A consumed by the `for` loop, which gets assigned to the `c` variable and results in the output `--> A`.

❽ Iteration continues with a second call to `next(g)`, advancing the generator body from `yield 'A'` to `yield 'B'`. The text `continue` is output by the second `print` in the generator body.

❾ `yield 'B'` yields the value B consumed by the `for` loop, which gets assigned to the `c` loop variable, so the loop prints `--> B`.

❿ Iteration continues with a third call to `next(it)`, advancing to the end of the body of the function. The text `end.` appears in the output because of the third `print` in the generator body.

⓫ When the generator function runs to the end, the generator object raises `StopIt eration`. The `for` loop machinery catches that exception, and the loop terminates cleanly.

Now hopefully it's clear how `Sentence.__iter__` in Example 17-5 works: `__iter__` is a generator function which, when called, builds a generator object that implements the `Iterator` interface, so the `SentenceIterator` class is no longer needed.

That second version of `Sentence` is more concise than the first, but it's not as lazy as it could be. Nowadays, laziness is considered a good trait, at least in programming languages and APIs. A lazy implementation postpones producing values to the last possible moment. This saves memory and may avoid wasting CPU cycles, too.

We'll build lazy `Sentence` classes next.

Lazy Sentences

The final variations of `Sentence` are lazy, taking advantage of a lazy function from the `re` module.

Sentence Take #4: Lazy Generator

The `Iterator` interface is designed to be lazy: `next(my_iterator)` yields one item at a time. The opposite of lazy is eager: lazy evaluation and eager evaluation are technical terms in programming language theory.

Our Sentence implementations so far have not been lazy because the __init__ eagerly builds a list of all words in the text, binding it to the self.words attribute. This requires processing the entire text, and the list may use as much memory as the text itself (probably more; it depends on how many nonword characters are in the text). Most of this work will be in vain if the user only iterates over the first couple of words. If you wonder, "Is there a lazy way of doing this in Python?" the answer is often "Yes."

The re.finditer function is a lazy version of re.findall. Instead of a list, re.finditer returns a generator yielding re.MatchObject instances on demand. If there are many matches, re.finditer saves a lot of memory. Using it, our third version of Sentence is now lazy: it only reads the next word from the text when it is needed. The code is in Example 17-8.

Example 17-8. sentence_gen2.py: Sentence implemented using a generator function calling the re.finditer generator function

```python
import re
import reprlib

RE_WORD = re.compile(r'\w+')


class Sentence:

    def __init__(self, text):
        self.text = text  ❶

    def __repr__(self):
        return f'Sentence({reprlib.repr(self.text)})'

    def __iter__(self):
        for match in RE_WORD.finditer(self.text):  ❷
            yield match.group()  ❸
```

❶ No need to have a words list.

❷ finditer builds an iterator over the matches of RE_WORD on self.text, yielding MatchObject instances.

❸ match.group() extracts the matched text from the MatchObject instance.

Generators are a great shortcut, but the code can be made even more concise with a generator expression.

Sentence Take #5: Lazy Generator Expression

We can replace simple generator functions like the one in the previous Sentence class (Example 17-8) with a generator expression. As a list comprehension builds lists, a generator expression builds generator objects. Example 17-9 contrasts their behavior.

Example 17-9. The gen_AB generator function is used by a list comprehension, then by a generator expression

```
>>> def gen_AB():  ❶
...     print('start')
...     yield 'A'
...     print('continue')
...     yield 'B'
...     print('end.')
...
>>> res1 = [x*3 for x in gen_AB()]  ❷
start
continue
end.
>>> for i in res1:  ❸
...     print('-->', i)
...
--> AAA
--> BBB
>>> res2 = (x*3 for x in gen_AB())  ❹
>>> res2
<generator object <genexpr> at 0x10063c240>
>>> for i in res2:  ❺
...     print('-->', i)
...
start       ❻
--> AAA
continue
--> BBB
end.
```

❶ This is the same gen_AB function from Example 17-7.

❷ The list comprehension eagerly iterates over the items yielded by the generator object returned by gen_AB(): 'A' and 'B'. Note the output in the next lines: start, continue, end.

❸ This for loop iterates over the res1 list built by the list comprehension.

❹ The generator expression returns res2, a generator object. The generator is not consumed here.

❺ Only when the `for` loop iterates over `res2`, this generator gets items from `gen_AB`. Each iteration of the `for` loop implicitly calls `next(res2)`, which in turn calls `next()` on the generator object returned by `gen_AB()`, advancing it to the next `yield`.

❻ Note how the output of `gen_AB()` interleaves with the output of the `print` in the `for` loop.

We can use a generator expression to further reduce the code in the `Sentence` class. See Example 17-10.

Example 17-10. sentence_genexp.py: Sentence *implemented using a generator expression*

```
import re
import reprlib

RE_WORD = re.compile(r'\w+')


class Sentence:

    def __init__(self, text):
        self.text = text

    def __repr__(self):
        return f'Sentence({reprlib.repr(self.text)})'

    def __iter__(self):
        return (match.group() for match in RE_WORD.finditer(self.text))
```

The only difference from Example 17-8 is the `__iter__` method, which here is not a generator function (it has no `yield`) but uses a generator expression to build a generator and then returns it. The end result is the same: the caller of `__iter__` gets a generator object.

Generator expressions are syntactic sugar: they can always be replaced by generator functions, but sometimes are more convenient. The next section is about generator expression usage.

When to Use Generator Expressions

I used several generator expressions when implementing the `Vector` class in Example 12-16. Each of these methods has a generator expression: `__eq__`, `__hash__`, `__abs__`, `angle`, `angles`, `format`, `__add__`, and `__mul__`. In all those methods, a list comprehension would also work, at the cost of using more memory to store the intermediate list values.

In Example 17-10, we saw that a generator expression is a syntactic shortcut to create a generator without defining and calling a function. On the other hand, generator functions are more flexible: we can code complex logic with multiple statements, and we can even use them as *coroutines*, as we'll see in "Classic Coroutines" on page 645.

For the simpler cases, a generator expression is easier to read at a glance, as the Vector example shows.

My rule of thumb in choosing the syntax to use is simple: if the generator expression spans more than a couple of lines, I prefer to code a generator function for the sake of readability.

Syntax Tip

When a generator expression is passed as the single argument to a function or constructor, you don't need to write a set of parentheses for the function call and another to enclose the generator expression. A single pair will do, like in the Vector call from the __mul__ method in Example 12-16, reproduced here:

```
def __mul__(self, scalar):
    if isinstance(scalar, numbers.Real):
        return Vector(n * scalar for n in self)
    else:
        return NotImplemented
```

However, if there are more function arguments after the generator expression, you need to enclose it in parentheses to avoid a Syntax Error.

The Sentence examples we've seen demonstrate generators playing the role of the classic Iterator pattern: retrieving items from a collection. But we can also use generators to yield values independent of a data source. The next section shows an example.

But first, a short discussion on the overlapping concepts of *iterator* and *generator*.

Contrasting Iterators and Generators

In the official Python documentation and codebase, the terminology around iterators and generators is inconsistent and evolving. I've adopted the following definitions:

iterator
> General term for any object that implements a __next__ method. Iterators are designed to produce data that is consumed by the client code, i.e., the code that drives the iterator via a for loop or other iterative feature, or by explicitly calling

next(it) on the iterator—although this explicit usage is much less common. In practice, most iterators we use in Python are *generators*.

generator

An iterator built by the Python compiler. To create a generator, we don't implement __next__. Instead, we use the yield keyword to make a *generator function*, which is a factory of *generator objects*. A *generator expression* is another way to build a generator object. Generator objects provide __next__, so they are iterators. Since Python 3.5, we also have *asynchronous generators* declared with async def. We'll study them in Chapter 21, "Asynchronous Programming".

The *Python Glossary* (*https://fpy.li/17-8*) recently introduced the term *generator iterator* (*https://fpy.li/17-9*) to refer to generator objects built by generator functions, while the entry for *generator expression* (*https://fpy.li/17-10*) says it returns an "iterator."

But the objects returned in both cases are generator objects, according to Python:

```
>>> def g():
...     yield 0
...
>>> g()
<generator object g at 0x10e6fb290>
>>> ge = (c for c in 'XYZ')
>>> ge
<generator object <genexpr> at 0x10e936ce0>
>>> type(g()), type(ge)
(<class 'generator'>, <class 'generator'>)
```

An Arithmetic Progression Generator

The classic Iterator pattern is all about traversal: navigating some data structure. But a standard interface based on a method to fetch the next item in a series is also useful when the items are produced on the fly, instead of retrieved from a collection. For example, the range built-in generates a bounded arithmetic progression (AP) of integers. What if you need to generate an AP of numbers of any type, not only integers?

Example 17-11 shows a few console tests of an ArithmeticProgression class we will see in a moment. The signature of the constructor in Example 17-11 is Arithmetic Progression(begin, step[, end]). The complete signature of the range built-in is range(start, stop[, step]). I chose to implement a different signature because the step is mandatory but end is optional in an arithmetic progression. I also changed the argument names from start/stop to begin/end to make it clear that I opted for a different signature. In each test in Example 17-11, I call list() on the result to inspect the generated values.

Example 17-11. Demonstration of an ArithmeticProgression class

```
>>> ap = ArithmeticProgression(0, 1, 3)
>>> list(ap)
[0, 1, 2]
>>> ap = ArithmeticProgression(1, .5, 3)
>>> list(ap)
[1.0, 1.5, 2.0, 2.5]
>>> ap = ArithmeticProgression(0, 1/3, 1)
>>> list(ap)
[0.0, 0.3333333333333333, 0.6666666666666666]
>>> from fractions import Fraction
>>> ap = ArithmeticProgression(0, Fraction(1, 3), 1)
>>> list(ap)
[Fraction(0, 1), Fraction(1, 3), Fraction(2, 3)]
>>> from decimal import Decimal
>>> ap = ArithmeticProgression(0, Decimal('.1'), .3)
>>> list(ap)
[Decimal('0'), Decimal('0.1'), Decimal('0.2')]
```

Note that the type of the numbers in the resulting arithmetic progression follows the type of begin + step, according to the numeric coercion rules of Python arithmetic. In Example 17-11, you see lists of int, float, Fraction, and Decimal numbers. Example 17-12 lists the implementation of the ArithmeticProgression class.

Example 17-12. The ArithmeticProgression class

```
class ArithmeticProgression:

    def __init__(self, begin, step, end=None):       ❶
        self.begin = begin
        self.step = step
        self.end = end   # None -> "infinite" series

    def __iter__(self):
        result_type = type(self.begin + self.step)   ❷
        result = result_type(self.begin)             ❸
        forever = self.end is None                   ❹
        index = 0
        while forever or result < self.end:          ❺
            yield result                             ❻
            index += 1
            result = self.begin + self.step * index  ❼
```

❶ __init__ requires two arguments: begin and step; end is optional, if it's None, the series will be unbounded.

❷ Get the type of adding self.begin and self.step. For example, if one is int and the other is float, result_type will be float.

❸ This line makes a `result` with the same numeric value of `self.begin`, but coerced to the type of the subsequent additions.[7]

❹ For readability, the `forever` flag will be `True` if the `self.end` attribute is `None`, resulting in an unbounded series.

❺ This loop runs `forever` or until the result matches or exceeds `self.end`. When this loop exits, so does the function.

❻ The current `result` is produced.

❼ The next potential result is calculated. It may never be yielded, because the `while` loop may terminate.

In the last line of Example 17-12, instead of adding `self.step` to the previous `result` each time around the loop, I opted to ignore the previous `result` and each new `result` by adding `self.begin` to `self.step` multiplied by `index`. This avoids the cumulative effect of floating-point errors after successive additions. These simple experiments make the difference clear:

```
>>> 100 * 1.1
110.00000000000001
>>> sum(1.1 for _ in range(100))
109.99999999999982
>>> 1000 * 1.1
1100.0
>>> sum(1.1 for _ in range(1000))
1100.0000000000086
```

The `ArithmeticProgression` class from Example 17-12 works as intended, and is a another example of using a generator function to implement the `__iter__` special method. However, if the whole point of a class is to build a generator by implementing `__iter__`, we can replace the class with a generator function. A generator function is, after all, a generator factory.

---

7 In Python 2, there was a `coerce()` built-in function, but it's gone in Python 3. It was deemed unnecessary because the numeric coercion rules are implicit in the arithmetic operator methods. So the best way I could think of to coerce the initial value to be of the same type as the rest of the series was to perform the addition and use its type to convert the result. I asked about this in the Python-list and got an excellent response from Steven D'Aprano (*https://fpy.li/17-11*).

Example 17-13 shows a generator function called `aritprog_gen` that does the same job as `ArithmeticProgression` but with less code. The tests in Example 17-11 all pass if you just call `aritprog_gen` instead of `ArithmeticProgression`.[8]

Example 17-13. The `aritprog_gen` generator function

```
def aritprog_gen(begin, step, end=None):
    result = type(begin + step)(begin)
    forever = end is None
    index = 0
    while forever or result < end:
        yield result
        index += 1
        result = begin + step * index
```

Example 17-13 is elegant, but always remember: there are plenty of ready-to-use generators in the standard library, and the next section will show a shorter implementation using the `itertools` module.

Arithmetic Progression with itertools

The `itertools` module in Python 3.10 has 20 generator functions that can be combined in a variety of interesting ways.

For example, the `itertools.count` function returns a generator that yields numbers. Without arguments, it yields a series of integers starting with 0. But you can provide optional `start` and `step` values to achieve a result similar to our `aritprog_gen` functions:

```
>>> import itertools
>>> gen = itertools.count(1, .5)
>>> next(gen)
1
>>> next(gen)
1.5
>>> next(gen)
2.0
>>> next(gen)
2.5
```

---

8 The *17-it-generator/* directory in the *Fluent Python* code repository (*https://fpy.li/code*) includes doctests and a script, *aritprog_runner.py*, which runs the tests against all variations of the *aritprog*.py* scripts.

`itertools.count` never stops, so if you call `list(count())`, Python will try to build a `list` that would fill all the memory chips ever made. In practice, your machine will become very grumpy long before the call fails.

On the other hand, there is the `itertools.takewhile` function: it returns a generator that consumes another generator and stops when a given predicate evaluates to `False`. So we can combine the two and write this:

```
>>> gen = itertools.takewhile(lambda n: n < 3, itertools.count(1, .5))
>>> list(gen)
[1, 1.5, 2.0, 2.5]
```

Leveraging `takewhile` and `count`, Example 17-14 is even more concise.

Example 17-14. aritprog_v3.py: this works like the previous aritprog_gen functions

```
import itertools


def aritprog_gen(begin, step, end=None):
    first = type(begin + step)(begin)
    ap_gen = itertools.count(first, step)
    if end is None:
        return ap_gen
    return itertools.takewhile(lambda n: n < end, ap_gen)
```

Note that `aritprog_gen` in Example 17-14 is not a generator function: it has no `yield` in its body. But it returns a generator, just as a generator function does.

However, recall that `itertools.count` adds the `step` repeatedly, so the floating-point series it produces are not as precise as Example 17-13.

The point of Example 17-14 is: when implementing generators, know what is available in the standard library, otherwise there's a good chance you'll reinvent the wheel. That's why the next section covers several ready-to-use generator functions.

Generator Functions in the Standard Library

The standard library provides many generators, from plain-text file objects providing line-by-line iteration, to the awesome `os.walk` (*https://fpy.li/17-12*) function, which yields filenames while traversing a directory tree, making recursive filesystem searches as simple as a `for` loop.

The `os.walk` generator function is impressive, but in this section I want to focus on general-purpose functions that take arbitrary iterables as arguments and return generators that yield selected, computed, or rearranged items. In the following tables, I summarize two dozen of them, from the built-in, `itertools`, and `functools` modules. For convenience, I grouped them by high-level functionality, regardless of where they are defined.

The first group contains the filtering generator functions: they yield a subset of items produced by the input iterable, without changing the items themselves. Like `take while`, most functions listed in Table 17-1 take a `predicate`, which is a one-argument Boolean function that will be applied to each item in the input to determine whether the item is included in the output.

Table 17-1. Filtering generator functions

Module	Function	Description
itertools	compress(it, selector_it)	Consumes two iterables in parallel; yields items from `it` whenever the corresponding item in `selector_it` is truthy
itertools	dropwhile(predicate, it)	Consumes `it`, skipping items while `predicate` computes truthy, then yields every remaining item (no further checks are made)
(built-in)	filter(predicate, it)	Applies `predicate` to each item of `iterable`, yielding the item if `predicate(item)` is truthy; if `predicate` is None, only truthy items are yielded
itertools	filterfalse(predicate, it)	Same as `filter`, with the `predicate` logic negated: yields items whenever `predicate` computes falsy
itertools	islice(it, stop) or islice(it, start, stop, step=1)	Yields items from a slice of `it`, similar to `s[:stop]` or `s[start:stop:step]` except `it` can be any iterable, and the operation is lazy
itertools	takewhile(predicate, it)	Yields items while `predicate` computes truthy, then stops and no further checks are made

The console listing in Example 17-15 shows the use of all the functions in Table 17-1.

Example 17-15. Filtering generator functions examples

```
>>> def vowel(c):
...     return c.lower() in 'aeiou'
...
>>> list(filter(vowel, 'Aardvark'))
['A', 'a', 'a']
>>> import itertools
>>> list(itertools.filterfalse(vowel, 'Aardvark'))
['r', 'd', 'v', 'r', 'k']
>>> list(itertools.dropwhile(vowel, 'Aardvark'))
['r', 'd', 'v', 'a', 'r', 'k']
>>> list(itertools.takewhile(vowel, 'Aardvark'))
['A', 'a']
>>> list(itertools.compress('Aardvark', (1, 0, 1, 1, 0, 1)))
```

```
['A', 'r', 'd', 'a']
>>> list(itertools.islice('Aardvark', 4))
['A', 'a', 'r', 'd']
>>> list(itertools.islice('Aardvark', 4, 7))
['v', 'a', 'r']
>>> list(itertools.islice('Aardvark', 1, 7, 2))
['a', 'd', 'a']
```

The next group contains the mapping generators: they yield items computed from each individual item in the input iterable—or iterables, in the case of map and star map.[9] The generators in Table 17-2 yield one result per item in the input iterables. If the input comes from more than one iterable, the output stops as soon as the first input iterable is exhausted.

Table 17-2. Mapping generator functions

Module	Function	Description
itertools	accumulate(it, [func])	Yields accumulated sums; if func is provided, yields the result of applying it to the first pair of items, then to the first result and next item, etc.
(built-in)	enumerate(iterable, start=0)	Yields 2-tuples of the form (index, item), where index is counted from start, and item is taken from the iterable
(built-in)	map(func, it1, [it2, …, itN])	Applies func to each item of it, yielding the result; if N iterables are given, func must take N arguments and the iterables will be consumed in parallel
itertools	starmap(func, it)	Applies func to each item of it, yielding the result; the input iterable should yield iterable items iit, and func is applied as func(*iit)

Example 17-16 demonstrates some uses of itertools.accumulate.

Example 17-16. itertools.accumulate generator function examples

```
>>> sample = [5, 4, 2, 8, 7, 6, 3, 0, 9, 1]
>>> import itertools
>>> list(itertools.accumulate(sample))   ❶
[5, 9, 11, 19, 26, 32, 35, 35, 44, 45]
>>> list(itertools.accumulate(sample, min))   ❷
[5, 4, 2, 2, 2, 2, 2, 0, 0, 0]
>>> list(itertools.accumulate(sample, max))   ❸
[5, 5, 5, 8, 8, 8, 8, 8, 9, 9]
>>> import operator
>>> list(itertools.accumulate(sample, operator.mul))   ❹
[5, 20, 40, 320, 2240, 13440, 40320, 0, 0, 0]
```

---

9 Here, the term "mapping" is unrelated to dictionaries, but has to do with the map built-in.

```
>>> list(itertools.accumulate(range(1, 11), operator.mul))
[1, 2, 6, 24, 120, 720, 5040, 40320, 362880, 3628800]  ❺
```

❶ Running sum.

❷ Running minimum.

❸ Running maximum.

❹ Running product.

❺ Factorials from 1! to 10!.

The remaining functions of Table 17-2 are shown in Example 17-17.

Example 17-17. Mapping generator function examples

```
>>> list(enumerate('albatroz', 1))  ❶
[(1, 'a'), (2, 'l'), (3, 'b'), (4, 'a'), (5, 't'), (6, 'r'), (7, 'o'), (8, 'z')]
>>> import operator
>>> list(map(operator.mul, range(11), range(11)))  ❷
[0, 1, 4, 9, 16, 25, 36, 49, 64, 81, 100]
>>> list(map(operator.mul, range(11), [2, 4, 8]))  ❸
[0, 4, 16]
>>> list(map(lambda a, b: (a, b), range(11), [2, 4, 8]))  ❹
[(0, 2), (1, 4), (2, 8)]
>>> import itertools
>>> list(itertools.starmap(operator.mul, enumerate('albatroz', 1)))  ❺
['a', 'll', 'bbb', 'aaaa', 'ttttt', 'rrrrrr', 'ooooooo', 'zzzzzzzz']
>>> sample = [5, 4, 2, 8, 7, 6, 3, 0, 9, 1]
>>> list(itertools.starmap(lambda a, b: b / a,
...         enumerate(itertools.accumulate(sample), 1)))  ❻
[5.0, 4.5, 3.6666666666666665, 4.75, 5.2, 5.333333333333333,
5.0, 4.375, 4.888888888888889, 4.5]
```

❶ Number the letters in the word, starting from 1.

❷ Squares of integers from 0 to 10.

❸ Multiplying numbers from two iterables in parallel: results stop when the shortest iterable ends.

❹ This is what the zip built-in function does.

❺ Repeat each letter in the word according to its place in it, starting from 1.

❻ Running average.

Next, we have the group of merging generators—all of these yield items from multiple input iterables. `chain` and `chain.from_iterable` consume the input iterables sequentially (one after the other), while `product`, `zip`, and `zip_longest` consume the input iterables in parallel. See Table 17-3.

Table 17-3. Generator functions that merge multiple input iterables

Module	Function	Description
itertools	chain(it1, ..., itN)	Yields all items from it1, then from it2, etc., seamlessly
itertools	chain.from_iterable(it)	Yields all items from each iterable produced by it, one after the other, seamlessly; it will be an iterable where the items are also iterables, for example, a list of tuples
itertools	product(it1, ..., itN, repeat=1)	Cartesian product: yields N-tuples made by combining items from each input iterable, like nested for loops could produce; repeat allows the input iterables to be consumed more than once
(built-in)	zip(it1, ..., itN, strict=False)	Yields N-tuples built from items taken from the iterables in parallel, silently stopping when the first iterable is exhausted, unless strict=True is given[a]
itertools	zip_longest(it1, ..., itN, fillvalue=None)	Yields N-tuples built from items taken from the iterables in parallel, stopping only when the last iterable is exhausted, filling the blanks with the fillvalue

[a] The strict keyword-only argument is new in Python 3.10. When strict=True, ValueError is raised if any iterable has a different length. The default is False, for backward compatibility.

Example 17-18 shows the use of the `itertools.chain` and `zip` generator functions and their siblings. Recall that the `zip` function is named after the zip fastener or zipper (no relation to compression). Both `zip` and `itertools.zip_longest` were introduced in "The Awesome zip" on page 418.

Example 17-18. Merging generator function examples

```
>>> list(itertools.chain('ABC', range(2)))  ❶
['A', 'B', 'C', 0, 1]
>>> list(itertools.chain(enumerate('ABC')))  ❷
[(0, 'A'), (1, 'B'), (2, 'C')]
>>> list(itertools.chain.from_iterable(enumerate('ABC')))  ❸
[0, 'A', 1, 'B', 2, 'C']
>>> list(zip('ABC', range(5), [10, 20, 30, 40]))  ❹
[('A', 0, 10), ('B', 1, 20), ('C', 2, 30)]
>>> list(itertools.zip_longest('ABC', range(5)))  ❺
[('A', 0), ('B', 1), ('C', 2), (None, 3), (None, 4)]
>>> list(itertools.zip_longest('ABC', range(5), fillvalue='?'))  ❻
[('A', 0), ('B', 1), ('C', 2), ('?', 3), ('?', 4)]
```

❶ chain is usually called with two or more iterables.

❷ chain does nothing useful when called with a single iterable.

❸ But chain.from_iterable takes each item from the iterable, and chains them in sequence, as long as each item is itself iterable.

❹ Any number of iterables can be consumed by zip in parallel, but the generator always stops as soon as the first iterable ends. In Python ≥ 3.10, if the strict=True argument is given and an iterable ends before the others, ValueError is raised.

❺ itertools.zip_longest works like zip, except it consumes all input iterables to the end, padding output tuples with None, as needed.

❻ The fillvalue keyword argument specifies a custom padding value.

The itertools.product generator is a lazy way of computing Cartesian products, which we built using list comprehensions with more than one for clause in "Cartesian Products" on page 27. Generator expressions with multiple for clauses can also be used to produce Cartesian products lazily. Example 17-19 demonstrates itertools.product.

Example 17-19. itertools.product generator function examples

```
>>> list(itertools.product('ABC', range(2)))  ❶
[('A', 0), ('A', 1), ('B', 0), ('B', 1), ('C', 0), ('C', 1)]
>>> suits = 'spades hearts diamonds clubs'.split()
>>> list(itertools.product('AK', suits))  ❷
[('A', 'spades'), ('A', 'hearts'), ('A', 'diamonds'), ('A', 'clubs'),
 ('K', 'spades'), ('K', 'hearts'), ('K', 'diamonds'), ('K', 'clubs')]
>>> list(itertools.product('ABC'))  ❸
[('A',), ('B',), ('C',)]
>>> list(itertools.product('ABC', repeat=2))  ❹
[('A', 'A'), ('A', 'B'), ('A', 'C'), ('B', 'A'), ('B', 'B'),
 ('B', 'C'), ('C', 'A'), ('C', 'B'), ('C', 'C')]
>>> list(itertools.product(range(2), repeat=3))
[(0, 0, 0), (0, 0, 1), (0, 1, 0), (0, 1, 1), (1, 0, 0),
 (1, 0, 1), (1, 1, 0), (1, 1, 1)]
>>> rows = itertools.product('AB', range(2), repeat=2)
>>> for row in rows: print(row)
...
('A', 0, 'A', 0)
('A', 0, 'A', 1)
('A', 0, 'B', 0)
('A', 0, 'B', 1)
('A', 1, 'A', 0)
```

```
('A', 1, 'A', 1)
('A', 1, 'B', 0)
('A', 1, 'B', 1)
('B', 0, 'A', 0)
('B', 0, 'A', 1)
('B', 0, 'B', 0)
('B', 0, 'B', 1)
('B', 1, 'A', 0)
('B', 1, 'A', 1)
('B', 1, 'B', 0)
('B', 1, 'B', 1)
```

❶ The Cartesian product of a str with three characters and a range with two integers yields six tuples (because 3 * 2 is 6).

❷ The product of two card ranks ('AK') and four suits is a series of eight tuples.

❸ Given a single iterable, product yields a series of one-tuples—not very useful.

❹ The repeat=N keyword argument tells the product to consume each input iterable N times.

Some generator functions expand the input by yielding more than one value per input item. They are listed in Table 17-4.

Table 17-4. Generator functions that expand each input item into multiple output items

Module	Function	Description
itertools	combinations(it, out_len)	Yields combinations of out_len items from the items yielded by it
itertools	combinations_with_replacement(it, out_len)	Yields combinations of out_len items from the items yielded by it, including combinations with repeated items
itertools	count(start=0, step=1)	Yields numbers starting at start, incremented by step, indefinitely
itertools	cycle(it)	Yields items from it, storing a copy of each, then yields the entire sequence repeatedly, indefinitely
itertools	pairwise(it)	Yields successive overlapping pairs taken from the input iterable[a]
itertools	permutations(it, out_len=None)	Yields permutations of out_len items from the items yielded by it; by default, out_len is len(list(it))
itertools	repeat(item, [times])	Yields the given item repeatedly, indefinitely unless a number of times is given

[a] itertools.pairwise was added in Python 3.10.

The count and repeat functions from itertools return generators that conjure items out of nothing: neither of them takes an iterable as input. We saw iter tools.count in "Arithmetic Progression with itertools" on page 622. The cycle generator makes a backup of the input iterable and yields its items repeatedly. Example 17-20 illustrates the use of count, cycle, pairwise, and repeat.

Example 17-20. count, cycle, pairwise, and repeat

```
>>> ct = itertools.count()  ❶
>>> next(ct)  ❷
0
>>> next(ct), next(ct), next(ct)  ❸
(1, 2, 3)
>>> list(itertools.islice(itertools.count(1, .3), 3))  ❹
[1, 1.3, 1.6]
>>> cy = itertools.cycle('ABC')  ❺
>>> next(cy)
'A'
>>> list(itertools.islice(cy, 7))  ❻
['B', 'C', 'A', 'B', 'C', 'A', 'B']
>>> list(itertools.pairwise(range(7)))  ❼
[(0, 1), (1, 2), (2, 3), (3, 4), (4, 5), (5, 6)]
>>> rp = itertools.repeat(7)  ❽
>>> next(rp), next(rp)
(7, 7)
>>> list(itertools.repeat(8, 4))  ❾
[8, 8, 8, 8]
>>> list(map(operator.mul, range(11), itertools.repeat(5)))  ❿
[0, 5, 10, 15, 20, 25, 30, 35, 40, 45, 50]
```

❶ Build a count generator ct.

❷ Retrieve the first item from ct.

❸ I can't build a list from ct, because ct never stops, so I fetch the next three items.

❹ I can build a list from a count generator if it is limited by islice or takewhile.

❺ Build a cycle generator from 'ABC' and fetch its first item, 'A'.

❻ A list can only be built if limited by islice; the next seven items are retrieved here.

❼ For each item in the input, pairwise yields a 2-tuple with that item and the next —if there is a next item. Available in Python ≥ 3.10.

❽ Build a `repeat` generator that will yield the number 7 forever.

❾ A `repeat` generator can be limited by passing the `times` argument: here the number 8 will be produced 4 times.

❿ A common use of `repeat`: providing a fixed argument in `map`; here it provides the 5 multiplier.

The `combinations`, `combinations_with_replacement`, and `permutations` generator functions—together with `product`—are called the *combinatorics generators* in the `itertools` documentation page (*https://fpy.li/17-13*). There is a close relationship between `itertools.product` and the remaining *combinatoric* functions as well, as Example 17-21 shows.

Example 17-21. Combinatoric generator functions yield multiple values per input item

```
>>> list(itertools.combinations('ABC', 2))  ❶
[('A', 'B'), ('A', 'C'), ('B', 'C')]
>>> list(itertools.combinations_with_replacement('ABC', 2))  ❷
[('A', 'A'), ('A', 'B'), ('A', 'C'), ('B', 'B'), ('B', 'C'), ('C', 'C')]
>>> list(itertools.permutations('ABC', 2))  ❸
[('A', 'B'), ('A', 'C'), ('B', 'A'), ('B', 'C'), ('C', 'A'), ('C', 'B')]
>>> list(itertools.product('ABC', repeat=2))  ❹
[('A', 'A'), ('A', 'B'), ('A', 'C'), ('B', 'A'), ('B', 'B'), ('B', 'C'),
('C', 'A'), ('C', 'B'), ('C', 'C')]
```

❶ All combinations of `len()==2` from the items in `'ABC'`; item ordering in the generated tuples is irrelevant (they could be sets).

❷ All combinations of `len()==2` from the items in `'ABC'`, including combinations with repeated items.

❸ All permutations of `len()==2` from the items in `'ABC'`; item ordering in the generated tuples is relevant.

❹ Cartesian product from `'ABC'` and `'ABC'` (that's the effect of `repeat=2`).

The last group of generator functions we'll cover in this section are designed to yield all items in the input iterables, but rearranged in some way. Here are two functions that return multiple generators: `itertools.groupby` and `itertools.tee`. The other generator function in this group, the `reversed` built-in, is the only one covered in this section that does not accept any iterable as input, but only sequences. This makes sense: because `reversed` will yield the items from last to first, it only works with a sequence with a known length. But it avoids the cost of making a reversed copy of the sequence by yielding each item as needed. I put the `itertools.product` function together with the *merging* generators in Table 17-3 because they all consume more than one iterable, while the generators in Table 17-5 all accept at most one input iterable.

Table 17-5. Rearranging generator functions

Module	Function	Description
itertools	groupby(it, key=None)	Yields 2-tuples of the form (key, group), where key is the grouping criterion and group is a generator yielding the items in the group
(built-in)	reversed(seq)	Yields items from seq in reverse order, from last to first; seq must be a sequence or implement the __reversed__ special method
itertools	tee(it, n=2)	Yields a tuple of *n* generators, each yielding the items of the input iterable independently

Example 17-22 demonstrates the use of `itertools.groupby` and the `reversed` built-in. Note that `itertools.groupby` assumes that the input iterable is sorted by the grouping criterion, or at least that the items are clustered by that criterion—even if not completely sorted. Tech reviewer Miroslav Šedivý suggested this use case: you can sort the `datetime` objects chronologically, then `groupby` weekday to get a group of Monday data, followed by Tuesday data, etc., and then by Monday (of the next week) again, and so on.

Example 17-22. `itertools.groupby`

```
>>> list(itertools.groupby('LLLLAAGGG'))  ❶
[('L', <itertools._grouper object at 0x102227cc0>),
 ('A', <itertools._grouper object at 0x102227b38>),
 ('G', <itertools._grouper object at 0x102227b70>)]
>>> for char, group in itertools.groupby('LLLLAAAGG'):  ❷
...     print(char, '->', list(group))
...
L -> ['L', 'L', 'L', 'L']
A -> ['A', 'A',]
G -> ['G', 'G', 'G']
>>> animals = ['duck', 'eagle', 'rat', 'giraffe', 'bear',
...            'bat', 'dolphin', 'shark', 'lion']
>>> animals.sort(key=len)  ❸
>>> animals
['rat', 'bat', 'duck', 'bear', 'lion', 'eagle', 'shark',
 'giraffe', 'dolphin']
>>> for length, group in itertools.groupby(animals, len):  ❹
...     print(length, '->', list(group))
...
3 -> ['rat', 'bat']
4 -> ['duck', 'bear', 'lion']
5 -> ['eagle', 'shark']
```

```
7 -> ['giraffe', 'dolphin']
>>> for length, group in itertools.groupby(reversed(animals), len): ❺
...     print(length, '->', list(group))
...
7 -> ['dolphin', 'giraffe']
5 -> ['shark', 'eagle']
4 -> ['lion', 'bear', 'duck']
3 -> ['bat', 'rat']
>>>
```

❶ groupby yields tuples of (key, group_generator).

❷ Handling groupby generators involves nested iteration: in this case, the outer for
    loop and the inner list constructor.

❸ Sort animals by length.

❹ Again, loop over the key and group pair, to display the key and expand the group
    into a list.

❺ Here the reverse generator iterates over animals from right to left.

The last of the generator functions in this group is iterator.tee, which has a unique
behavior: it yields multiple generators from a single input iterable, each yielding
every item from the input. Those generators can be consumed independently, as
shown in Example 17-23.

*Example 17-23. itertools.tee yields multiple generators, each yielding every item of
the input generator*

```
>>> list(itertools.tee('ABC'))
[<itertools._tee object at 0x10222abc8>, <itertools._tee object at 0x10222ac08>]
>>> g1, g2 = itertools.tee('ABC')
>>> next(g1)
'A'
>>> next(g2)
'A'
>>> next(g2)
'B'
>>> list(g1)
['B', 'C']
>>> list(g2)
['C']
>>> list(zip(*itertools.tee('ABC')))
[('A', 'A'), ('B', 'B'), ('C', 'C')]
```

Note that several examples in this section used combinations of generator functions. This is a great feature of these functions: because they take generators as arguments and return generators, they can be combined in many different ways.

Now we'll review another group of iterable-savvy functions in the standard library.

Iterable Reducing Functions

The functions in Table 17-6 all take an iterable and return a single result. They are known as "reducing," "folding," or "accumulating" functions. We can implement every one of the built-ins listed here with functools.reduce, but they exist as built-ins because they address some common use cases more easily. A longer explanation about functools.reduce appeared in "Vector Take #4: Hashing and a Faster ==" on page 413.

In the case of all and any, there is an important optimization functools.reduce does not support: all and any short-circuit—i.e., they stop consuming the iterator as soon as the result is determined. See the last test with any in Example 17-24.

Table 17-6. Built-in functions that read iterables and return single values

Module	Function	Description
(built-in)	all(it)	Returns True if all items in it are truthy, otherwise False; all([]) returns True
(built-in)	any(it)	Returns True if any item in it is truthy, otherwise False; any([]) returns False
(built-in)	max(it, [key=,] [default=])	Returns the maximum value of the items in it;[a] key is an ordering function, as in sorted; default is returned if the iterable is empty
(built-in)	min(it, [key=,] [default=])	Returns the minimum value of the items in it.[b] key is an ordering function, as in sorted; default is returned if the iterable is empty
functools	reduce(func, it, [initial])	Returns the result of applying func to the first pair of items, then to that result and the third item, and so on; if given, initial forms the initial pair with the first item
(built-in)	sum(it, start=0)	The sum of all items in it, with the optional start value added (use math.fsum for better precision when adding floats)

[a] May also be called as max(arg1, arg2, …, [key=?]), in which case the maximum among the arguments is returned.
[b] May also be called as min(arg1, arg2, …, [key=?]), in which case the minimum among the arguments is returned.

The operation of all and any is exemplified in Example 17-24.

Example 17-24. Results of all and any for some sequences

```
>>> all([1, 2, 3])
True
>>> all([1, 0, 3])
False
>>> all([])
True
>>> any([1, 2, 3])
True
>>> any([1, 0, 3])
True
>>> any([0, 0.0])
False
>>> any([])
False
>>> g = (n for n in [0, 0.0, 7, 8])
>>> any(g)  ❶
True
>>> next(g)  ❷
8
```

❶ any iterated over g until g yielded 7; then any stopped and returned True.

❷ That's why 8 was still remaining.

Another built-in that takes an iterable and returns something else is sorted. Unlike reversed, which is a generator function, sorted builds and returns a new list. After all, every single item of the input iterable must be read so they can be sorted, and the sorting happens in a list, therefore sorted just returns that list after it's done. I mention sorted here because it does consume an arbitrary iterable.

Of course, sorted and the reducing functions only work with iterables that eventually stop. Otherwise, they will keep on collecting items and never return a result.

If you've gotten this far, you've seen the most important and useful content of this chapter. The remaining sections cover advanced generator features that most of us don't see or need very often, such as the yield from construct and classic coroutines.

There are also sections about type hinting iterables, iterators, and classic coroutines.

The yield from syntax provides a new way of combining generators. That's next.

Subgenerators with yield from

The `yield from` expression syntax was introduced in Python 3.3 to allow a generator to delegate work to a subgenerator.

Before `yield from` was introduced, we used a `for` loop when a generator needed to yield values produced from another generator:

```
>>> def sub_gen():
...     yield 1.1
...     yield 1.2
...
>>> def gen():
...     yield 1
...     for i in sub_gen():
...         yield i
...     yield 2
...
>>> for x in gen():
...     print(x)
...
1
1.1
1.2
2
```

We can get the same result using `yield from`, as you can see in Example 17-25.

Example 17-25. Test-driving `yield from`

```
>>> def sub_gen():
...     yield 1.1
...     yield 1.2
...
>>> def gen():
...     yield 1
...     yield from sub_gen()
...     yield 2
...
>>> for x in gen():
...     print(x)
...
1
1.1
1.2
2
```

In Example 17-25, the for loop is the *client code*, gen is the *delegating generator*, and sub_gen is the *subgenerator*. Note that `yield from` pauses gen, and sub_gen takes over until it is exhausted. The values yielded by sub_gen pass through gen directly to the client for loop. Meanwhile, gen is suspended and cannot see the values passing through it. Only when sub_gen is done, gen resumes.

When the subgenerator contains a return statement with a value, that value can be captured in the delegating generator by using yield from as part of an expression. Example 17-26 demonstrates.

Example 17-26. yield from gets the return value of the subgenerator

```
>>> def sub_gen():
...     yield 1.1
...     yield 1.2
...     return 'Done!'
...
>>> def gen():
...     yield 1
...     result = yield from sub_gen()
...     print('<--', result)
...     yield 2
...
>>> for x in gen():
...     print(x)
...
1
1.1
1.2
<-- Done!
2
```

Now that we've seen the basics of yield from, let's study a couple of simple but practical examples of its use.

Reinventing chain

We saw in Table 17-3 that itertools provides a chain generator that yields items from several iterables, iterating over the first, then the second, and so on up to the last. This is a homemade implementation of chain with nested for loops in Python:[10]

```
>>> def chain(*iterables):
...     for it in iterables:
...         for i in it:
...             yield i
...
>>> s = 'ABC'
>>> r = range(3)
```

---

10 chain and most itertools functions are written in C.

```
>>> list(chain(s, r))
['A', 'B', 'C', 0, 1, 2]
```

The chain generator in the preceding code is delegating to each iterable it in turn, by driving each it in the inner for loop. That inner loop can be replaced with a yield from expression, as shown in the next console listing:

```
>>> def chain(*iterables):
...     for i in iterables:
...         yield from i
...
>>> list(chain(s, t))
['A', 'B', 'C', 0, 1, 2]
```

The use of yield from in this example is correct, and the code reads better, but it seems like syntactic sugar with little real gain. Now let's develop a more interesting example.

Traversing a Tree

In this section, we'll see yield from in a script to traverse a tree structure. I will build it in baby steps.

The tree structure for this example is Python's exception hierarchy (*https://fpy.li/ 17-14*). But the pattern can be adapted to show a directory tree or any other tree structure.

Starting from BaseException at level zero, the exception hierarchy is five levels deep as of Python 3.10. Our first baby step is to show level zero.

Given a root class, the tree generator in Example 17-27 yields its name and stops.

Example 17-27. tree/step0/tree.py: yield the name of the root class and stop

```
def tree(cls):
    yield cls.__name__


def display(cls):
    for cls_name in tree(cls):
        print(cls_name)


if __name__ == '__main__':
    display(BaseException)
```

The output of Example 17-27 is just one line:

```
BaseException
```

The next baby step takes us to level 1. The `tree` generator will yield the name of the root class and the names of each direct subclass. The names of the subclasses are indented to reveal the hierarchy. This is the output we want:

```
$ python3 tree.py
BaseException
    Exception
    GeneratorExit
    SystemExit
    KeyboardInterrupt
```

Example 17-28 produces that output.

Example 17-28. tree/step1/tree.py: yield the name of root class and direct subclasses

```python
def tree(cls):
    yield cls.__name__, 0                    ❶
    for sub_cls in cls.__subclasses__():     ❷
        yield sub_cls.__name__, 1            ❸


def display(cls):
    for cls_name, level in tree(cls):
        indent = ' ' * 4 * level             ❹
        print(f'{indent}{cls_name}')


if __name__ == '__main__':
    display(BaseException)
```

❶ To support the indented output, yield the name of the class and its level in the hierarchy.

❷ Use the `__subclasses__` special method to get a list of subclasses.

❸ Yield name of subclass and level 1.

❹ Build indentation string of 4 spaces times `level`. At level zero, this will be an empty string.

In Example 17-29, I refactor `tree` to separate the special case of the root class from the subclasses, which are now handled in the `sub_tree` generator. At `yield from`, the `tree` generator is suspended, and `sub_tree` takes over yielding values.

Example 17-29. tree/step2/tree.py: tree yields the root class name, then delegates to sub_tree

```
def tree(cls):
    yield cls.__name__, 0
    yield from sub_tree(cls)                ❶


def sub_tree(cls):
    for sub_cls in cls.__subclasses__():
        yield sub_cls.__name__, 1           ❷


def display(cls):
    for cls_name, level in tree(cls):       ❸
        indent = ' ' * 4 * level
        print(f'{indent}{cls_name}')


if __name__ == '__main__':
    display(BaseException)
```

❶ Delegate to sub_tree to yield the names of the subclasses.

❷ Yield the name of each subclass and level 1. Because of the yield from sub_tree(cls) inside tree, these values bypass the tree generator function completely...

❸ ... and are received directly here.

In keeping with the baby steps method, I'll write the simplest code I can imagine to reach level 2. For depth-first (*https://fpy.li/17-15*) tree traversal, after yielding each node in level 1, I want to yield the children of that node in level 2, before resuming level 1. A nested for loop takes care of that, as in Example 17-30.

Example 17-30. tree/step3/tree.py: sub_tree traverses levels 1 and 2 depth-first

```
def tree(cls):
    yield cls.__name__, 0
    yield from sub_tree(cls)


def sub_tree(cls):
    for sub_cls in cls.__subclasses__():
        yield sub_cls.__name__, 1
        for sub_sub_cls in sub_cls.__subclasses__():
            yield sub_sub_cls.__name__, 2
```

```
def display(cls):
    for cls_name, level in tree(cls):
        indent = ' ' * 4 * level
        print(f'{indent}{cls_name}')


if __name__ == '__main__':
    display(BaseException)
```

This is the result of running *step3/tree.py* from Example 17-30:

```
$ python3 tree.py
BaseException
    Exception
        TypeError
        StopAsyncIteration
        StopIteration
        ImportError
        OSError
        EOFError
        RuntimeError
        NameError
        AttributeError
        SyntaxError
        LookupError
        ValueError
        AssertionError
        ArithmeticError
        SystemError
        ReferenceError
        MemoryError
        BufferError
        Warning
    GeneratorExit
    SystemExit
    KeyboardInterrupt
```

You may already know where this is going, but I will stick to baby steps one more time: let's reach level 3 by adding yet another nested for loop. The rest of the program is unchanged, so Example 17-31 shows only the sub_tree generator.

Example 17-31. sub_tree generator from tree/step4/tree.py

```
def sub_tree(cls):
    for sub_cls in cls.__subclasses__():
        yield sub_cls.__name__, 1
        for sub_sub_cls in sub_cls.__subclasses__():
            yield sub_sub_cls.__name__, 2
            for sub_sub_sub_cls in sub_sub_cls.__subclasses__():
                yield sub_sub_sub_cls.__name__, 3
```

There is a clear pattern in Example 17-31. We do a for loop to get the subclasses of level N. Each time around the loop, we yield a subclass of level N, then start another for loop to visit level $N+1$.

In "Reinventing chain" on page 637, we saw how we can replace a nested for loop driving a generator with yield from on the same generator. We can apply that idea here, if we make sub_tree accept a level parameter, and yield from it recursively, passing the current subclass as the new root class with the next level number. See Example 17-32.

Example 17-32. tree/step5/tree.py: recursive sub_tree goes as far as memory allows

```
def tree(cls):
    yield cls.__name__, 0
    yield from sub_tree(cls, 1)


def sub_tree(cls, level):
    for sub_cls in cls.__subclasses__():
        yield sub_cls.__name__, level
        yield from sub_tree(sub_cls, level+1)


def display(cls):
    for cls_name, level in tree(cls):
        indent = ' ' * 4 * level
        print(f'{indent}{cls_name}')


if __name__ == '__main__':
    display(BaseException)
```

Example 17-32 can traverse trees of any depth, limited only by Python's recursion limit. The default limit allows 1,000 pending functions.

Any good tutorial about recursion will stress the importance of having a base case to avoid infinite recursion. A base case is a conditional branch that returns without making a recursive call. The base case is often implemented with an if statement. In Example 17-32, sub_tree has no if, but there is an implicit conditional in the for loop: if cls.__subclasses__() returns an empty list, the body of the loop is not executed, therefore no recursive call happens. The base case is when the cls class has no subclasses. In that case, sub_tree yields nothing. It just returns.

Example 17-32 works as intended, but we can make it more concise by recalling the pattern we observed when we reached level 3 (Example 17-31): we yield a subclass with level N, then start a nested for loop to visit level $N+1$. In Example 17-32 we replaced that nested loop with `yield from`. Now we can merge `tree` and `sub_tree` into a single generator. Example 17-33 is the last step for this example.

Example 17-33. tree/step6/tree.py: recursive calls of `tree` pass an incremented `level`
argument

```python
def tree(cls, level=0):
    yield cls.__name__, level
    for sub_cls in cls.__subclasses__():
        yield from tree(sub_cls, level+1)


def display(cls):
    for cls_name, level in tree(cls):
        indent = ' ' * 4 * level
        print(f'{indent}{cls_name}')


if __name__ == '__main__':
    display(BaseException)
```

At the start of "Subgenerators with yield from" on page 636, we saw how `yield from` connects the subgenerator directly to the client code, bypassing the delegating generator. That connection becomes really important when generators are used as coroutines and not only produce but also consume values from the client code, as we'll see in "Classic Coroutines" on page 645.

After this first encounter with `yield from`, let's turn to type hinting iterables and iterators.

Generic Iterable Types

Python's standard library has many functions that accept iterable arguments. In your code, such functions can be annotated like the `zip_replace` function we saw in Example 8-15, using `collections.abc.Iterable` (or `typing.Iterable` if you must support Python 3.8 or earlier, as explained in "Legacy Support and Deprecated Collection Types" on page 272). See Example 17-34.

Example 17-34. replacer.py returns an iterator of tuples of strings

```python
from collections.abc import Iterable

FromTo = tuple[str, str]  ❶

def zip_replace(text: str, changes: Iterable[FromTo]) -> str:  ❷
    for from_, to in changes:
```

```
        text = text.replace(from_, to)
    return text
```

❶ Define type alias; not required, but makes the next type hint more readable. Starting with Python 3.10, FromTo should have a type hint of typing.TypeAlias to clarify the reason for this line: FromTo: TypeAlias = tuple[str, str].

❷ Annotate changes to accept an Iterable of FromTo tuples.

Iterator types don't appear as often as Iterable types, but they are also simple to write. Example 17-35 shows the familiar Fibonacci generator, annotated.

Example 17-35. fibo_gen.py: fibonacci returns a generator of integers

```
from collections.abc import Iterator

def fibonacci() -> Iterator[int]:
    a, b = 0, 1
    while True:
        yield a
        a, b = b, a + b
```

Note that the type Iterator is used for generators coded as functions with yield, as well as iterators written "by hand" as classes with __next__. There is also a collec tions.abc.Generator type (and the corresponding deprecated typing.Generator) that we can use to annotate generator objects, but it is needlessly verbose for generators used as iterators.

Example 17-36, when checked with Mypy, reveals that the Iterator type is really a simplified special case of the Generator type.

Example 17-36. itergentype.py: two ways to annotate iterators

```
from collections.abc import Iterator
from keyword import kwlist
from typing import TYPE_CHECKING

short_kw = (k for k in kwlist if len(k) < 5)  ❶

if TYPE_CHECKING:
    reveal_type(short_kw)  ❷

long_kw: Iterator[str] = (k for k in kwlist if len(k) >= 4)  ❸

if TYPE_CHECKING:  ❹
    reveal_type(long_kw)
```

❶ Generator expression that yields Python keywords with less than 5 characters.

❷ Mypy infers: `typing.Generator[builtins.str*, None, None]`.[11]

❸ This also yields strings, but I added an explicit type hint.

❹ Revealed type: `typing.Iterator[builtins.str]`.

`abc.Iterator[str]` is *consistent-with* `abc.Generator[str, None, None]`, therefore Mypy issues no errors for type checking in Example 17-36.

`Iterator[T]` is a shortcut for `Generator[T, None, None]`. Both annotations mean "a generator that yields items of type `T`, but that does not consume or return values." Generators able to consume and return values are coroutines, our next topic.

Classic Coroutines

PEP 342—Coroutines via Enhanced Generators (*https://fpy.li/ pep342*) introduced the `.send()` and other features that made it possible to use generators as coroutines. PEP 342 uses the word "coroutine" with the same meaning I am using here.

It is unfortunate that Python's official documentation and standard library now use inconsistent terminology to refer to generators used as coroutines, forcing me to adopt the "classic coroutine" qualifier to contrast with the newer "native coroutine" objects.

After Python 3.5 came out, the trend is to use "coroutine" as a synonym for "native coroutine." But PEP 342 is not deprecated, and classic coroutines still work as originally designed, although they are no longer supported by `asyncio`.

Understanding classic coroutines in Python is confusing because they are actually generators used in a different way. So let's step back and consider another feature of Python that can be used in two ways.

We saw in "Tuples Are Not Just Immutable Lists" on page 30 that we can use `tuple` instances as records or as immutable sequences. When used as a record, a tuple is expected to have a specific number of items, and each item may have a different type. When used as immutable lists, a tuple can have any length, and all items are expected to have the same type. That's why there are two different ways to annotate tuples with type hints:

---

11 As of version 0.910, Mypy still uses the deprecated `typing` types.

```
# A city record with name, country, and population:
city: tuple[str, str, int]

# An immutable sequence of domain names:
domains: tuple[str, ...]
```

Something similar happens with generators. They are commonly used as iterators, but they can also be used as coroutines. A *coroutine* is really a generator function, created with the yield keyword in its body. And a *coroutine object* is physically a generator object. Despite sharing the same underlying implementation in C, the use cases of generators and coroutines in Python are so different that there are two ways to type hint them:

```
# The `readings` variable can be bound to an iterator
# or generator object that yields `float` items:
readings: Iterator[float]

# The `sim_taxi` variable can be bound to a coroutine
# representing a taxi cab in a discrete event simulation.
# It yields events, receives `float` timestamps, and returns
# the number of trips made during the simulation:
sim_taxi: Generator[Event, float, int]
```

Adding to the confusion, the typing module authors decided to name that type Gen erator, when in fact it describes the API of a generator object intended to be used as a coroutine, while generators are more often used as simple iterators.

The typing documentation (*https://fpy.li/17-17*) describes the formal type parameters of Generator like this:

```
Generator[YieldType, SendType, ReturnType]
```

The SendType is only relevant when the generator is used as a coroutine. That type parameter is the type of x in the call gen.send(x). It is an error to call .send() on a generator that was coded to behave as an iterator instead of a coroutine. Likewise, ReturnType is only meaningful to annotate a coroutine, because iterators don't return values like regular functions. The only sensible operation on a generator used as an iterator is to call next(it) directly or indirectly via for loops and other forms of iteration. The YieldType is the type of the value returned by a call to next(it).

The Generator type has the same type parameters as typing.Coroutine (*https://fpy.li/typecoro*):

```
Coroutine[YieldType, SendType, ReturnType]
```

The typing.Coroutine documentation (*https://fpy.li/typecoro*) actually says: "The variance and order of type variables correspond to those of Generator." But typ ing.Coroutine (deprecated) and collections.abc.Coroutine (generic since Python 3.9) are intended to annotate only native coroutines, not classic coroutines. If you want to use type hints with classic coroutines, you'll suffer through the confusion of annotating them as `Generator[YieldType, SendType, ReturnType]`.

David Beazley created some of the best talks and most comprehensive workshops about classic coroutines. In his PyCon 2009 course handout (*https://fpy.li/17-18*), he has a slide titled "Keeping It Straight," which reads:

- Generators produce data for iteration

- Coroutines are consumers of data

- To keep your brain from exploding, don't mix the two concepts together

- Coroutines are not related to iteration

- Note: There is a use of having `yield` produce a value in a coroutine, but it's not tied to iteration.[12]

Now let's see how classic coroutines work.

Example: Coroutine to Compute a Running Average

While discussing closures in Chapter 9, we studied objects to compute a running average. Example 9-7 shows a class and Example 9-13 presents a higher-order function returning a function that keeps the `total` and `count` variables across invocations in a closure. Example 17-37 shows how to do the same with a coroutine.[13]

Example 17-37. coroaverager.py: coroutine to compute a running average

```
from collections.abc import Generator

def averager() -> Generator[float, float, None]:  ❶
    total = 0.0
    count = 0
    average = 0.0
    while True:  ❷
        term = yield average  ❸
        total += term
        count += 1
        average = total/count
```

---

12 Slide 33, "Keeping It Straight," in "A Curious Course on Coroutines and Concurrency" (*https://fpy.li/17-18*).

13 This example is inspired by a snippet from Jacob Holm in the Python-ideas list, message titled "Yield-From: Finalization guarantees" (*https://fpy.li/17-20*). Some variations appear later in the thread, and Holm further explains his thinking in message 003912 (*https://fpy.li/17-21*).

❶ This function returns a generator that yields `float` values, accepts `float` values via `.send()`, and does not return a useful value.[14]

❷ This infinite loop means the coroutine will keep on yielding averages as long as the client code sends values.

❸ The `yield` statement here suspends the coroutine, yields a result to the client, and—later—gets a value sent by the caller to the coroutine, starting another iteration of the infinite loop.

In a coroutine, `total` and `count` can be local variables: no instance attributes or closures are needed to keep the context while the coroutine is suspended waiting for the next `.send()`. That's why coroutines are attractive replacements for callbacks in asynchronous programming—they keep local state between activations.

Example 17-38 runs doctests to show the `averager` coroutine in operation.

Example 17-38. coroaverager.py: doctest for the running average coroutine in Example 17-37

```
>>> coro_avg = averager()   ❶
>>> next(coro_avg)   ❷
0.0
>>> coro_avg.send(10)   ❸
10.0
>>> coro_avg.send(30)
20.0
>>> coro_avg.send(5)
15.0
```

❶ Create the coroutine object.

❷ Start the coroutine. This yields the initial value of `average`: 0.0.

❸ Now we are in business: each call to `.send()` yields the current average.

In Example 17-38, the call `next(coro_avg)` makes the coroutine advance to the `yield`, yielding the initial value for `average`. You can also start the coroutine by calling `coro_avg.send(None)`—this is actually what the `next()` built-in does. But you can't send any value other than `None`, because the coroutine can only accept a sent

---

14 In fact, it never returns unless some exception breaks the loop. Mypy 0.910 accepts both `None` and `typing`
   `.NoReturn` as the generator return type parameter—but it also accepts `str` in that position, so apparently it
   can't fully analyze the coroutine code at this time.

value when it is suspended at a yield line. Calling next() or .send(None) to advance to the first yield is known as "priming the coroutine."

After each activation, the coroutine is suspended precisely at the yield keyword, waiting for a value to be sent. The line coro_avg.send(10) provides that value, causing the coroutine to activate. The yield expression resolves to the value 10, assigning it to the term variable. The rest of the loop updates the total, count, and average variables. The next iteration in the while loop yields the average, and the coroutine is again suspended at the yield keyword.

The attentive reader may be anxious to know how the execution of an averager instance (e.g., coro_avg) may be terminated, because its body is an infinite loop. We don't usually need to terminate a generator, because it is garbage collected as soon as there are no more valid references to it. If you need to explicitly terminate it, use the .close() method, as shown in Example 17-39.

Example 17-39. coroaverager.py: continuing from Example 17-38

```
>>> coro_avg.send(20)    ❶
16.25
>>> coro_avg.close()     ❷
>>> coro_avg.close()     ❸
>>> coro_avg.send(5)     ❹
Traceback (most recent call last):
  ...
StopIteration
```

❶ coro_avg is the instance created in Example 17-38.

❷ The .close() method raises GeneratorExit at the suspended yield expression. If not handled in the coroutine function, the exception terminates it. Generator Exit is caught by the generator object that wraps the coroutine—that's why we don't see it.

❸ Calling .close() on a previously closed coroutine has no effect.

❹ Trying .send() on a closed coroutine raises StopIteration.

Besides the .send() method, PEP 342—Coroutines via Enhanced Generators (*https://fpy.li/pep342*) also introduced a way for a coroutine to return a value. The next section shows how.

Returning a Value from a Coroutine

We'll now study another coroutine to compute an average. This version will not yield partial results. Instead, it returns a tuple with the number of terms and the average. I've split the listing in two parts: Example 17-40 and Example 17-41.

Example 17-40. coroaverager2.py: top of the file

```
from collections.abc import Generator
from typing import Union, NamedTuple

class Result(NamedTuple):  ❶
    count: int  # type: ignore  ❷
    average: float

class Sentinel:  ❸
    def __repr__(self):
        return f'<Sentinel>'

STOP = Sentinel()  ❹

SendType = Union[float, Sentinel]  ❺
```

❶ The averager2 coroutine in Example 17-41 will return an instance of Result.

❷ The Result is actually a subclass of tuple, which has a .count() method that I don't need. The # type: ignore comment prevents Mypy from complaining about having a count field.[15]

❸ A class to make a sentinel value with a readable __repr__.

❹ The sentinel value that I'll use to make the coroutine stop collecting data and return a result.

❺ I'll use this type alias for the second type parameter of the coroutine Generator return type, the SendType parameter.

The SendType definition also works in Python 3.10, but if you don't need to support earlier versions, it is better to write it like this, after importing TypeAlias from typing:

---

15 I considered renaming the field, but count is the best name for the local variable in the coroutine, and is the name I used for this variable in similar examples in the book, so it makes sense to use the same name in the Result field. I don't hesitate to use # type: ignore to avoid the limitations and annoyances of static type checkers when submission to the tool would make the code worse or needlessly complicated.

```
SendType: TypeAlias = float | Sentinel
```

Using | instead of typing.Union is so concise and readable that I'd probably not create that type alias, but instead I'd write the signature of averager2 like this:

```
def averager2(verbose: bool=False) -> Generator[None, float | Sentinel, Result]:
```

Now, let's study the coroutine code itself (Example 17-41).

Example 17-41. coroaverager2.py: a coroutine that returns a result value

```
def averager2(verbose: bool = False) -> Generator[None, SendType, Result]:  ❶
    total = 0.0
    count = 0
    average = 0.0
    while True:
        term = yield  ❷
        if verbose:
            print('received:', term)
        if isinstance(term, Sentinel):  ❸
            break
        total += term  ❹
        count += 1
        average = total / count
    return Result(count, average)  ❺
```

❶ For this coroutine, the yield type is None because it does not yield data. It receives data of the SendType and returns a Result tuple when done.

❷ Using yield like this only makes sense in coroutines, which are designed to consume data. This yields None, but receives a term from .send(term).

❸ If the term is a Sentinel, break from the loop. Thanks to this isinstance check…

❹ …Mypy allows me to add term to the total without flagging an error that I can't add a float to an object that may be a float or a Sentinel.

❺ This line will be reached only if a Sentinel is sent to the coroutine.

Now let's see how we can use this coroutine, starting with a simple example that doesn't actually produce a result (Example 17-42).

Example 17-42. coroaverager2.py: doctest showing .cancel()

```
>>> coro_avg = averager2()
>>> next(coro_avg)
>>> coro_avg.send(10)  ❶
```

```
>>> coro_avg.send(30)
>>> coro_avg.send(6.5)
>>> coro_avg.close()  ❷
```

❶ Recall that averager2 does not yield partial results. It yields None, which Python's console omits.

❷ Calling .close() in this coroutine makes it stop but does not return a result, because the GeneratorExit exception is raised at the yield line in the coroutine, so the return statement is never reached.

Now let's make it work in Example 17-43.

Example 17-43. coroaverager2.py: doctest showing StopIteration with a Result

```
>>> coro_avg = averager2()
>>> next(coro_avg)
>>> coro_avg.send(10)
>>> coro_avg.send(30)
>>> coro_avg.send(6.5)
>>> try:
...     coro_avg.send(STOP)  ❶
... except StopIteration as exc:
...     result = exc.value  ❷
...
>>> result  ❸
Result(count=3, average=15.5)
```

❶ Sending the STOP sentinel makes the coroutine break from the loop and return a Result. The generator object that wraps the coroutine then raises StopIteration.

❷ The StopIteration instance has a value attribute bound to the value of the return statement that terminated the coroutine.

❸ Believe it or not!

This idea of "smuggling" the return value out of the coroutine wrapped in a StopIteration exception is a bizarre hack. Nevertheless, this bizarre hack is part of PEP 342— Coroutines via Enhanced Generators (*https://fpy.li/pep342*), and is documented with the StopIteration exception (*https://fpy.li/17-22*), and in the "Yield expressions" (*https://fpy.li/17-23*) section of Chapter 6 of *The Python Language Reference* (*https://fpy.li/17-24*).

A delegating generator can get the return value of a coroutine directly using the yield from syntax, as shown in Example 17-44.

Example 17-44. coroaverager2.py: doctest showing StopIteration with a Result

```
>>> def compute():
...     res = yield from averager2(True)   ❶
...     print('computed:', res)   ❷
...     return res   ❸
...
>>> comp = compute()   ❹
>>> for v in [None, 10, 20, 30, STOP]:   ❺
...     try:
...         comp.send(v)   ❻
...     except StopIteration as exc:   ❼
...         result = exc.value
received: 10
received: 20
received: 30
received: <Sentinel>
computed: Result(count=3, average=20.0)
>>> result   ❽
Result(count=3, average=20.0)
```

❶ res will collect the return value of averager2; the yield from machinery retrieves the return value when it handles the StopIteration exception that marks the termination of the coroutine. When True, the verbose parameter makes the coroutine print the value received, to make its operation visible.

❷ Keep an eye out for the output of this line when this generator runs.

❸ Return the result. This will also be wrapped in StopIteration.

❹ Create the delegating coroutine object.

❺ This loop will drive the delegating coroutine.

❻ First value sent is None, to prime the coroutine; last is the sentinel to stop it.

❼ Catch StopIteration to fetch the return value of compute.

❽ After the lines output by averager2 and compute, we get the Result instance.

Even though the examples here don't do much, the code is hard to follow. Driving the coroutine with .send() calls and retrieving results is complicated, except with yield from—but we can only use that syntax inside a delegating generator/coroutine, which must ultimately be driven by some nontrivial code, as shown in Example 17-44.

The previous examples show that using coroutines directly is cumbersome and confusing. Add exception handling and the coroutine .throw() method, and examples become even more convoluted. I won't cover .throw() in this book because—like .send()—it is only useful to drive coroutines "by hand," but I don't recommend doing that, unless you are creating a new coroutine-based framework from scratch.

 If you are interested in deeper coverage of classic coroutines—including the .throw() method—please check out "Classic Coroutines" (*https://fpy.li/oldcoro*) at the *fluentpython.com* companion website. That post includes Python-like pseudocode detailing how yield from drives generators and coroutines, as well as a a small discrete event simulation demonstrating a form of concurrency using coroutines without an asynchronous programming framework.

In practice, productive work with coroutines requires the support of a specialized framework. That is what asyncio provided for classic coroutines way back in Python 3.3. With the advent of native coroutines in Python 3.5, the Python core developers are gradually phasing out support for classic coroutines in asyncio. But the underlying mechanisms are very similar. The async def syntax makes native coroutines easier to spot in code, which is a great benefit. Inside, native coroutines use await instead of yield from to delegate to other coroutines. Chapter 21 is all about that.

Now let's wrap up the chapter with a mind-bending section about covariance and contravariance in type hints for coroutines.

Generic Type Hints for Classic Coroutines

Back in "Contravariant types" on page 554, I mentioned typing.Generator as one of the few standard library types with a contravariant type parameter. Now that we've studied classic coroutines, we are ready to make sense of this generic type.

Here is how typing.Generator was declared (*https://fpy.li/17-25*) in the *typing.py* module of Python 3.6:[16]

```
T_co = TypeVar('T_co', covariant=True)
V_co = TypeVar('V_co', covariant=True)
T_contra = TypeVar('T_contra', contravariant=True)

# many lines omitted
```

---

16 Since Python 3.7, typing.Generator and other types that correspond to ABCs in collections.abc were refactored with a wrapper around the corresponding ABC, so their generic parameters aren't visible in the *typing.py* source file. That's why I refer to Python 3.6 source code here.

```
class Generator(Iterator[T_co], Generic[T_co, T_contra, V_co],
                extra=_G_base):
```

That generic type declaration means that a Generator type hint requires those three type parameters we've seen before:

```
my_coro : Generator[YieldType, SendType, ReturnType]
```

From the type variables in the formal parameters, we see that YieldType and Return Type are covariant, but SendType is contravariant. To understand why, consider that YieldType and ReturnType are "output" types. Both describe data that comes out of the coroutine object—i.e., the generator object when used as a coroutine object.

It makes sense that these are covariant, because any code expecting a coroutine that yields floats can use a coroutine that yields integers. That's why Generator is covariant on its YieldType parameter. The same reasoning applies to the ReturnType parameter—also covariant.

Using the notation introduced in "Covariant types" on page 554, the covariance of the first and third parameters is expressed by the :> symbols pointing in the same direction:

```
               float :> int
Generator[float, Any, float] :> Generator[int, Any, int]
```

YieldType and ReturnType are examples of the first rule of "Variance rules of thumb" on page 555:

1. If a formal type parameter defines a type for data that comes out of the object, it can be covariant.

On the other hand, SendType is an "input" parameter: it is the type of the value argument for the .send(value) method of the coroutine object. Client code that needs to send floats to a coroutine cannot use a coroutine with int as the SendType because float is not a subtype of int. In other words, float is not *consistent-with* int. But the client can use a coroutine with complex as the SendType, because float is a subtype of complex, therefore float is *consistent-with* complex.

The :> notation makes the contravariance of the second parameter visible:

```
               float :> int
Generator[Any, float, Any] <: Generator[Any, int, Any]
```

This is an example of the second Variance Rule of Thumb:

2. If a formal type parameter defines a type for data that goes into the object after its initial construction, it can be contravariant.

This merry discussion of variance completes the longest chapter in the book.

Chapter Summary

Iteration is so deeply embedded in the language that I like to say that Python groks iterators.[17] The integration of the Iterator pattern in the semantics of Python is a prime example of how design patterns are not equally applicable in all programming languages. In Python, a classic Iterator implemented "by hand" as in Example 17-4 has no practical use, except as a didactic example.

In this chapter, we built a few versions of a class to iterate over individual words in text files that may be very long. We saw how Python uses the iter() built-in to create iterators from sequence-like objects. We build a classic iterator as a class with __next__(), and then we used generators to make each successive refactoring of the Sentence class more concise and readable.

We then coded a generator of arithmetic progressions and showed how to leverage the itertools module to make it simpler. An overview of most general-purpose generator functions in the standard library followed.

We then studied yield from expressions in the context of simple generators with the chain and tree examples.

The last major section was about classic coroutines, a topic of waning importance after native coroutines were added in Python 3.5. Although difficult to use in practice, classic coroutines are the foundation of native coroutines, and the yield from expression is the direct precursor of await.

Also covered were type hints for Iterable, Iterator, and Generator types—with the latter providing a concrete and rare example of a contravariant type parameter.

Further Reading

A detailed technical explanation of generators appears in *The Python Language Reference* in "6.2.9. Yield expressions" (*https://fpy.li/17-27*). The PEP where generator functions were defined is PEP 255—Simple Generators (*https://fpy.li/pep255*).

The itertools module documentation (*https://fpy.li/17-28*) is excellent because of all the examples included. Although the functions in that module are implemented in C, the documentation shows how some of them would be written in Python, often by leveraging other functions in the module. The usage examples are also great; for instance, there is a snippet showing how to use the accumulate function to amortize a loan with interest, given a list of payments over time. There is also an "Itertools

---

17 According to the Jargon file (*https://fpy.li/17-26*), to *grok* is not merely to learn something, but to absorb it so "it becomes part of you, part of your identity."

Recipes" (*https://fpy.li/17-29*) section with additional high-performance functions that use the `itertools` functions as building blocks.

Beyond Python's standard library, I recommend the More Itertools (*https://fpy.li/17-30*) package, which follows the fine `itertools` tradition in providing powerful generators with plenty of examples and some useful recipes.

Chapter 4, "Iterators and Generators," of *Python Cookbook*, 3rd ed., by David Beazley and Brian K. Jones (O'Reilly), has 16 recipes covering this subject from many different angles, focusing on practical applications. It includes some illuminating recipes with `yield from`.

Sebastian Rittau—currently a top contributor of *typeshed*—explains why iterators should be iterable, as he noted in 2006 that, "Java: Iterators are not Iterable" (*https://fpy.li/17-31*).

The `yield from` syntax is explained with examples in the "What's New in Python 3.3" section of PEP 380—Syntax for Delegating to a Subgenerator (*https://fpy.li/17-32*). My post "Classic Coroutines" (*https://fpy.li/oldcoro*) at *fluentpython.com* explains `yield from` in depth, including Python pseudocode of its implementation in C.

David Beazley is the ultimate authority on Python generators and coroutines. The *Python Cookbook*, 3rd ed., (O'Reilly) he coauthored with Brian Jones has numerous recipes with coroutines. Beazley's PyCon tutorials on the subject are famous for their depth and breadth. The first was at PyCon US 2008: "Generator Tricks for Systems Programmers" (*https://fpy.li/17-33*). PyCon US 2009 saw the legendary "A Curious Course on Coroutines and Concurrency" (*https://fpy.li/17-34*) (hard-to-find video links for all three parts: part 1 (*https://fpy.li/17-35*), part 2 (*https://fpy.li/17-36*), and part 3 (*https://fpy.li/17-37*)). His tutorial from PyCon 2014 in Montréal was "Generators: The Final Frontier" (*https://fpy.li/17-38*), in which he tackles more concurrency examples—so it's really more about topics in Chapter 21. Dave can't resist making brains explode in his classes, so in the last part of "The Final Frontier," coroutines replace the classic Visitor pattern in an arithmetic expression evaluator.

Coroutines allow new ways of organizing code, and just as recursion or polymorphism (dynamic dispatch), it takes some time getting used to their possibilities. An interesting example of classic algorithm rewritten with coroutines is in the post "Greedy algorithm with coroutines" (*https://fpy.li/17-39*), by James Powell.

Brett Slatkin's *Effective Python*, 1st ed. (*https://fpy.li/17-40*) (Addison-Wesley) has an excellent short chapter titled "Consider Coroutines to Run Many Functions Concurrently." That chapter is not in the second edition of *Effective Python*, but it is still available online as a sample chapter (*https://fpy.li/17-41*). Slatkin presents the best example of driving coroutines with `yield from` that I've seen: an implementation of John Conway's Game of Life (*https://fpy.li/17-42*) in which coroutines manage the state of each cell as the game runs. I refactored the code for the Game of Life example —separating the functions and classes that implement the game from the testing snippets used in Slatkin's original code. I also rewrote the tests as doctests, so you can see the output of the various coroutines and classes without running the script. The refactored example (*https://fpy.li/17-43*) is posted as a GitHub gist (*https://fpy.li/17-44*).

Soapbox

The Minimalistic Iterator Interface in Python

In the "Implementation" section of the Iterator pattern,[18] the Gang of Four wrote:

> The minimal interface to Iterator consists of the operations First, Next, IsDone, and CurrentItem.

However, that very sentence has a footnote that reads:

> We can make this interface even smaller by merging Next, IsDone, and CurrentItem into a single operation that advances to the next object and returns it. If the traversal is finished, then this operation returns a special value (0, for instance) that marks the end of the iteration.

This is close to what we have in Python: the single method __next__ does the job. But instead of using a sentinel, which could be overlooked by mistake, the StopIteration exception signals the end of the iteration. Simple and correct: that's the Python way.

Pluggable Generators

Anyone who manages large datasets finds many uses for generators. This is the story of the first time I built a practical solution around generators.

Years ago I worked at BIREME, a digital library run by PAHO/WHO (Pan-American Health Organization/World Health Organization) in São Paulo, Brazil. Among the bibliographic datasets created by BIREME are LILACS (Latin American and Caribbean Health Sciences index) and SciELO (Scientific Electronic Library Online), two comprehensive databases indexing the research literature about health sciences produced in the region.

Since the late 1980s, the database system used to manage LILACS is CDS/ISIS, a nonrelational document database created by UNESCO. One of my jobs was to research alternatives for a possible migration of LILACS—and eventually the much larger SciELO—to a modern, open source, document database such as CouchDB or MongoDB. At the time, I wrote a paper explaining the semistructured data model and

---

18 Gamma et. al., *Design Patterns: Elements of Reusable Object-Oriented Software*, p. 261.

different ways to represent CDS/ISIS data with JSON-like records: "From ISIS to CouchDB: Databases and Data Models for Bibliographic Records" (*https://fpy.li/ 17-45*).

As part of that research, I wrote a Python script to read a CDS/ISIS file and write a JSON file suitable for importing to CouchDB or MongoDB. Initially, the script read files in the ISO-2709 format exported by CDS/ISIS. The reading and writing had to be done incrementally because the full datasets were much bigger than main memory. That was easy enough: each iteration of the main for loop read one record from the *.iso* file, massaged it, and wrote it to the *.json* output.

However, for operational reasons, it was deemed necessary that *isis2json.py* supported another CDS/ISIS data format: the binary *.mst* files used in production at BIREME— to avoid the costly export to ISO-2709. Now I had a problem: the libraries used to read ISO-2709 and *.mst* files had very different APIs. And the JSON writing loop was already complicated because the script accepted a variety of command-line options to restructure each output record. Reading data using two different APIs in the same for loop where the JSON was produced would be unwieldy.

The solution was to isolate the reading logic into a pair of generator functions: one for each supported input format. In the end, I split the *isis2json.py* script into four functions. You can see the Python 2 source code with dependencies in the *fluentpy- thon/isis2json* (*https://fpy.li/17-46*) repository on GitHub.[19]

Here is a high-level overview of how the script is structured:

main

> The main function uses argparse to read command-line options that configure the structure of the output records. Based on the input filename extension, a suit- able generator function is selected to read the data and yield the records, one by one.

iter_iso_records

> This generator function reads *.iso* files (assumed to be in the ISO-2709 format). It takes two arguments: the filename and isis_json_type, one of the options related to the record structure. Each iteration of its for loop reads one record, creates an empty dict, populates it with field data, and yields the dict.

---

19 The code is in Python 2 because one of its optional dependencies is a Java library named *Bruma*, which we can import when we run the script with Jython—which does not yet support Python 3.

`iter_mst_records`

> This other generator functions reads *.mst* files.[20] If you look at the source code for *isis2json.py*, you'll see that it's not as simple as `iter_iso_records`, but its interface and overall structure is the same: it takes a filename and an `isis_json_type` argument and enters a `for` loop, which builds and yields one `dict` per iteration, representing a single record.

`write_json`

> This function performs the actual writing of the JSON records, one at a time. It takes numerous arguments, but the first one—`input_gen`—is a reference to a generator function: either `iter_iso_records` or `iter_mst_records`. The main `for` loop in `write_json` iterates over the dictionaries yielded by the selected `input_gen` generator, restructures it in different ways as determined by the command-line options, and appends the JSON record to the output file.

By leveraging generator functions, I was able to decouple the reading from the writing. Of course, the simplest way to decouple them would be to read all records to memory, then write them to disk. But that was not a viable option because of the size of the datasets. Using generators, the reading and writing is interleaved, so the script can process files of any size. Also, the special logic for reading a record in the different input formats is separated from the logic of restructuring each record for writing.

Now, if we need *isis2json.py* to support an additional input format—say, MARCXML, a DTD used by the US Library of Congress to represent ISO-2709 data—it will be easy to add a third generator function to implement the reading logic, without changing anything in the complicated `write_json` function.

This is not rocket science, but it's a real example where generators enabled an efficient and flexible solution to process databases as a stream of records, keeping memory usage low regardless of the size of the dataset.

---

20 The library used to read the complex *.mst* binary is actually written in Java, so this functionality is only available when *isis2json.py* is executed with the Jython interpreter, version 2.5 or newer. For further details, see the *README.rst* (*https://fpy.li/17-47*) file in the repository. The dependencies are imported inside the generator functions that need them, so the script can run even if only one of the external libraries is available.

with, match, and else Blocks

Context managers may end up being almost as important as the subroutine itself. We've only scratched the surface with them. [...] Basic has a with statement, there are with statements in lots of languages. But they don't do the same thing, they all do something very shallow, they save you from repeated dotted [attribute] lookups, they don't do setup and tear down. Just because it's the same name don't think it's the same thing. The with statement is a very big deal.[1]

—Raymond Hettinger, eloquent Python evangelist

This chapter is about control flow features that are not so common in other languages, and for this reason tend to be overlooked or underused in Python. They are:

- The with statement and context manager protocol
- Pattern matching with match/case
- The else clause in for, while, and try statements

The with statement sets up a temporary context and reliably tears it down, under the control of a context manager object. This prevents errors and reduces boilerplate code, making APIs at the same time safer and easier to use. Python programmers are finding lots of uses for with blocks beyond automatic file closing.

We've seen pattern matching in previous chapters, but here we'll see how the grammar of a language can be expressed as sequence patterns. That observation explains why match/case is an effective tool to create language processors that are easy to understand and extend. We'll study a complete interpreter for a small but functional

---

[1] PyCon US 2013 keynote: "What Makes Python Awesome" (*https://fpy.li/18-1*); the part about with starts at 23:00 and ends at 26:15.

subset of the Scheme language. The same ideas could be applied to develop a template language or a DSL (Domain-Specific Language) to encode business rules in a larger system.

The else clause is not a big deal, but it does help convey intention when properly used together with for, while, and try.

What's New in This Chapter

"Pattern Matching in lis.py: A Case Study" on page 673 is a new section.

I updated "The contextlib Utilities" on page 667 to cover a few features of the context lib module added since Python 3.6, and the new parenthesized context managers syntax introduced in Python 3.10.

Let's start with the powerful with statement.

Context Managers and with Blocks

Context manager objects exist to control a with statement, just like iterators exist to control a for statement.

The with statement was designed to simplify some common uses of try/finally, which guarantees that some operation is performed after a block of code, even if the block is terminated by return, an exception, or a sys.exit() call. The code in the finally clause usually releases a critical resource or restores some previous state that was temporarily changed.

The Python community is finding new, creative uses for context managers. Some examples from the standard library are:

- Managing transactions in the sqlite3 module—see "Using the connection as a context manager" (*https://fpy.li/18-2*).

- Safely handling locks, conditions, and semaphores—as described in the thread ing module documentation (*https://fpy.li/18-3*).

- Setting up custom environments for arithmetic operations with Decimal objects —see the decimal.localcontext documentation (*https://fpy.li/18-4*).

- Patching objects for testing—see the unittest.mock.patch function (*https://fpy.li/18-5*).

The context manager interface consists of the __enter__ and __exit__ methods. At the top of the with, Python calls the __enter__ method of the context manager object. When the with block completes or terminates for any reason, Python calls __exit__ on the context manager object.

The most common example is making sure a file object is closed. Example 18-1 is a detailed demonstration of using with to close a file.

Example 18-1. Demonstration of a file object as a context manager

```
>>> with open('mirror.py') as fp:  ❶
...     src = fp.read(60)  ❷
...
>>> len(src)
60
>>> fp  ❸
<_io.TextIOWrapper name='mirror.py' mode='r' encoding='UTF-8'>
>>> fp.closed, fp.encoding  ❹
(True, 'UTF-8')
>>> fp.read(60)  ❺
Traceback (most recent call last):
  File "<stdin>", line 1, in <module>
ValueError: I/O operation on closed file.
```

❶ fp is bound to the opened text file because the file's __enter__ method returns self.

❷ Read 60 Unicode characters from fp.

❸ The fp variable is still available—with blocks don't define a new scope, as functions do.

❹ We can read the attributes of the fp object.

❺ But we can't read more text from fp because at the end of the with block, the TextIOWrapper.__exit__ method was called, and it closed the file.

The first callout in Example 18-1 makes a subtle but crucial point: the context manager object is the result of evaluating the expression after with, but the value bound to the target variable (in the as clause) is the result returned by the __enter__ method of the context manager object.

It just happens that the open() function returns an instance of TextIOWrapper, and its __enter__ method returns self. But in a different class, the __enter__ method may also return some other object instead of the context manager instance.

When control flow exits the with block in any way, the __exit__ method is invoked on the context manager object, not on whatever was returned by __enter__.

The as clause of the with statement is optional. In the case of open, we always need it to get a reference to the file, so that we can call methods on it. But some context managers return None because they have no useful object to give back to the user.

Example 18-2 shows the operation of a perfectly frivolous context manager designed to highlight the distinction between the context manager and the object returned by its __enter__ method.

Example 18-2. Test-driving the LookingGlass context manager class

```
>>> from mirror import LookingGlass
>>> with LookingGlass() as what:  ❶
...      print('Alice, Kitty and Snowdrop')  ❷
...      print(what)
...
pordwonS dna yttiK ,ecilA
YKCOWREBBAJ
>>> what  ❸
'JABBERWOCKY'
>>> print('Back to normal.')  ❹
Back to normal.
```

❶ The context manager is an instance of LookingGlass; Python calls __enter__ on the context manager and the result is bound to what.

❷ Print a str, then the value of the target variable what. The output of each print will come out reversed.

❸ Now the with block is over. We can see that the value returned by __enter__, held in what, is the string 'JABBERWOCKY'.

❹ Program output is no longer reversed.

Example 18-3 shows the implementation of LookingGlass.

Example 18-3. mirror.py: code for the LookingGlass context manager class

```
import sys

class LookingGlass:

    def __enter__(self):  ❶
        self.original_write = sys.stdout.write  ❷
        sys.stdout.write = self.reverse_write  ❸
        return 'JABBERWOCKY'  ❹

    def reverse_write(self, text):  ❺
```

```
        self.original_write(text[::-1])

    def __exit__(self, exc_type, exc_value, traceback):  ❻
        sys.stdout.write = self.original_write  ❼
        if exc_type is ZeroDivisionError:  ❽
            print('Please DO NOT divide by zero!')
            return True  ❾
    ❿
```

❶ Python invokes __enter__ with no arguments besides self.

❷ Hold the original sys.stdout.write method, so we can restore it later.

❸ Monkey-patch sys.stdout.write, replacing it with our own method.

❹ Return the 'JABBERWOCKY' string just so we have something to put in the target variable what.

❺ Our replacement to sys.stdout.write reverses the text argument and calls the original implementation.

❻ Python calls __exit__ with None, None, None if all went well; if an exception is raised, the three arguments get the exception data, as described after this example.

❼ Restore the original method to sys.stdout.write.

❽ If the exception is not None and its type is ZeroDivisionError, print a message…

❾ …and return True to tell the interpreter that the exception was handled.

❿ If __exit__ returns None or any *falsy* value, any exception raised in the with block will be propagated.

 When real applications take over standard output, they often want to replace sys.stdout with another file-like object for a while, then switch back to the original. The contextlib.redirect_stdout (*https://fpy.li/18-6*) context manager does exactly that: just pass it the file-like object that will stand in for sys.stdout.

The interpreter calls the __enter__ method with no arguments—beyond the implicit self. The three arguments passed to __exit__ are:

`exc_type`
> The exception class (e.g., `ZeroDivisionError`).

`exc_value`
> The exception instance. Sometimes, parameters passed to the exception constructor—such as the error message—can be found in `exc_value.args`.

`traceback`
> A `traceback` object.[2]

For a detailed look at how a context manager works, see Example 18-4, where `LookingGlass` is used outside of a `with` block, so we can manually call its `__enter__` and `__exit__` methods.

Example 18-4. Exercising `LookingGlass` *without a* `with` *block*

```
>>> from mirror import LookingGlass
>>> manager = LookingGlass()     ❶
>>> manager  # doctest: +ELLIPSIS
<mirror.LookingGlass object at 0x...>
>>> monster = manager.__enter__()     ❷
>>> monster == 'JABBERWOCKY'     ❸
eurT
>>> monster
'YKCOWREBBAJ'
>>> manager  # doctest: +ELLIPSIS
>... ta tcejbo ssalGgnikooL.rorrim<
>>> manager.__exit__(None, None, None)     ❹
>>> monster
'JABBERWOCKY'
```

❶ Instantiate and inspect the `manager` instance.

❷ Call the manager's `__enter__` method and store result in `monster`.

❸ `monster` is the string `'JABBERWOCKY'`. The `True` identifier appears reversed because all output via `stdout` goes through the `write` method we patched in `__enter__`.

❹ Call `manager.__exit__` to restore the previous `stdout.write`.

---

2 The three arguments received by `self` are exactly what you get if you call `sys.exc_info()` (*https://fpy.li/18-7*) in the `finally` block of a `try/finally` statement. This makes sense, considering that the `with` statement is meant to replace most uses of `try/finally`, and calling `sys.exc_info()` was often necessary to determine what clean-up action would be required.

Parenthesized Context Managers in Python 3.10

Python 3.10 adopted a new, more powerful parser (*https://fpy.li/ pep617*), allowing new syntax beyond what was possible with the older LL(1) parser (*https://fpy.li/18-8*). One syntax enhancement was to allow parenthesized context managers, like this:

```
with (
    CtxManager1() as example1,
    CtxManager2() as example2,
    CtxManager3() as example3,
):
    ...
```

Prior to 3.10, we'd have to write that as nested with blocks.

The standard library includes the contextlib package with handy functions, classes, and decorators for building, combining, and using context managers.

The contextlib Utilities

Before rolling your own context manager classes, take a look at contextlib—"Utilities for with-statement contexts" (*https://fpy.li/18-9*) in the Python documentation. Maybe what you are about to build already exists, or there is a class or some callable that will make your job easier.

Besides the redirect_stdout context manager mentioned right after Example 18-3, redirect_stderr was added in Python 3.5—it does the same as the former, but for output directed to stderr.

The contextlib package also includes:

closing
    A function to build context managers out of objects that provide a close() method but don't implement the __enter__/__exit__ interface.

suppress
    A context manager to temporarily ignore exceptions given as arguments.

nullcontext
    A context manager that does nothing, to simplify conditional logic around objects that may not implement a suitable context manager. It serves as a stand-in when conditional code before the with block may or may not provide a context manager for the with statement—added in Python 3.7.

The contextlib module provides classes and a decorator that are more widely applicable than the decorators just mentioned:

`@contextmanager`
>A decorator that lets you build a context manager from a simple generator function, instead of creating a class and implementing the interface. See "Using @contextmanager" on page 668.

`AbstractContextManager`
>An ABC that formalizes the context manager interface, and makes it a bit easier to create context manager classes by subclassing—added in Python 3.6.

`ContextDecorator`
>A base class for defining class-based context managers that can also be used as function decorators, running the entire function within a managed context.

`ExitStack`
>A context manager that lets you enter a variable number of context managers. When the `with` block ends, `ExitStack` calls the stacked context managers' `__exit__` methods in LIFO order (last entered, first exited). Use this class when you don't know beforehand how many context managers you need to enter in your `with` block; for example, when opening all files from an arbitrary list of files at the same time.

With Python 3.7, `contextlib` added `AbstractAsyncContextManager`, `@asynccontext manager`, and `AsyncExitStack`. They are similar to the equivalent utilities without the `async` part of the name, but designed for use with the new `async with` statement, covered in Chapter 21.

The most widely used of these utilities is the `@contextmanager` decorator, so it deserves more attention. That decorator is also interesting because it shows a use for the `yield` statement unrelated to iteration.

Using @contextmanager

The `@contextmanager` decorator is an elegant and practical tool that brings together three distinctive Python features: a function decorator, a generator, and the `with` statement.

Using `@contextmanager` reduces the boilerplate of creating a context manager: instead of writing a whole class with `__enter__`/`__exit__` methods, you just implement a generator with a single `yield` that should produce whatever you want the `__enter__` method to return.

In a generator decorated with `@contextmanager`, `yield` splits the body of the function in two parts: everything before the `yield` will be executed at the beginning of the `with` block when the interpreter calls `__enter__`; the code after `yield` will run when `__exit__` is called at the end of the block.

Example 18-5 replaces the LookingGlass class from Example 18-3 with a generator function.

Example 18-5. mirror_gen.py: a context manager implemented with a generator

```
import contextlib
import sys

@contextlib.contextmanager   ❶
def looking_glass():
    original_write = sys.stdout.write   ❷

    def reverse_write(text):   ❸
        original_write(text[::-1])

    sys.stdout.write = reverse_write   ❹
    yield 'JABBERWOCKY'   ❺
    sys.stdout.write = original_write   ❻
```

❶ Apply the contextmanager decorator.

❷ Preserve the original sys.stdout.write method.

❸ reverse_write can call original_write later because it is available in its closure.

❹ Replace sys.stdout.write with reverse_write.

❺ Yield the value that will be bound to the target variable in the as clause of the with statement. The generator pauses at this point while the body of the with executes.

❻ When control exits the with block, execution continues after the yield; here the original sys.stdout.write is restored.

Example 18-6 shows the looking_glass function in operation.

Example 18-6. Test-driving the looking_glass context manager function

```
>>> from mirror_gen import looking_glass
>>> with looking_glass() as what:   ❶
...     print('Alice, Kitty and Snowdrop')
...     print(what)
...
pordwonS dna yttiK ,ecilA
YKCOWREBBAJ
>>> what
'JABBERWOCKY'
```

```
>>> print('back to normal')
back to normal
```

❶ The only difference from Example 18-2 is the name of the context manager: look
ing_glass instead of LookingGlass.

The contextlib.contextmanager decorator wraps the function in a class that implements the __enter__ and __exit__ methods.[3]

The __enter__ method of that class:

1. Calls the generator function to get a generator object—let's call it gen.
2. Calls next(gen) to drive it to the yield keyword.
3. Returns the value yielded by next(gen), to allow the user to bind it to a variable
   in the with/as form.

When the with block terminates, the __exit__ method:

1. Checks whether an exception was passed as exc_type; if so, gen.throw(excep
   tion) is invoked, causing the exception to be raised in the yield line inside the
   generator function body.
2. Otherwise, next(gen) is called, resuming the execution of the generator function
   body after the yield.

Example 18-5 has a flaw: if an exception is raised in the body of the with block, the
Python interpreter will catch it and raise it again in the yield expression inside look
ing_glass. But there is no error handling there, so the looking_glass generator will
terminate without ever restoring the original sys.stdout.write method, leaving the
system in an invalid state.

Example 18-7 adds special handling of the ZeroDivisionError exception, making it
functionally equivalent to the class-based Example 18-3.

*Example 18-7. mirror_gen_exc.py: generator-based context manager implementing
exception handling—same external behavior as Example 18-3*

```
import contextlib
import sys

@contextlib.contextmanager
```

---

3 The actual class is named _GeneratorContextManager. If you want to see exactly how it works, read its source
  code (*https://fpy.li/18-10*) in *Lib/contextlib.py* in Python 3.10.

```
def looking_glass():
    original_write = sys.stdout.write

    def reverse_write(text):
        original_write(text[::-1])

    sys.stdout.write = reverse_write
    msg = ''  ❶
    try:
        yield 'JABBERWOCKY'
    except ZeroDivisionError:  ❷
        msg = 'Please DO NOT divide by zero!'
    finally:
        sys.stdout.write = original_write  ❸
        if msg:
            print(msg)  ❹
```

❶ Create a variable for a possible error message; this is the first change in relation to Example 18-5.

❷ Handle ZeroDivisionError by setting an error message.

❸ Undo monkey-patching of sys.stdout.write.

❹ Display error message, if it was set.

Recall that the __exit__ method tells the interpreter that it has handled the exception by returning a truthy value; in that case, the interpreter suppresses the exception. On the other hand, if __exit__ does not explicitly return a value, the interpreter gets the usual None, and propagates the exception. With @contextmanager, the default behavior is inverted: the __exit__ method provided by the decorator assumes any exception sent into the generator is handled and should be suppressed.

Having a try/finally (or a with block) around the yield is an unavoidable price of using @contextmanager, because you never know what the users of your context manager are going to do inside the with block.[4]

---

4 This tip is quoted literally from a comment by Leonardo Rochael, one of the tech reviewers for this book. Nicely said, Leo!

A little-known feature of @contextmanager is that the generators decorated with it can also be used as decorators themselves.[5] That happens because @contextmanager is implemented with the contextlib.ContextDecorator class.

Example 18-8 shows the looking_glass context manager from Example 18-5 used as decorator.

Example 18-8. The looking_glass context manager also works as a decorator

```
>>> @looking_glass()
... def verse():
...     print('The time has come')
...
>>> verse()  ❶
emoc sah emit ehT
>>> print('back to normal')  ❷
back to normal
```

❶ looking_glass does its job before and after the body of verse runs.

❷ This confirms that the original sys.write was restored.

Contrast Example 18-8 with Example 18-6, where looking_glass is used as a context manager.

An interesting real-life example of @contextmanager outside of the standard library is Martijn Pieters' in-place file rewriting using a context manager (*https://fpy.li/18-11*). Example 18-9 shows how it's used.

Example 18-9. A context manager for rewriting files in place

```
import csv

with inplace(csvfilename, 'r', newline='') as (infh, outfh):
    reader = csv.reader(infh)
    writer = csv.writer(outfh)

    for row in reader:
        row += ['new', 'columns']
        writer.writerow(row)
```

The inplace function is a context manager that gives you two handles—infh and outfh in the example—to the same file, allowing your code to read and write to it at

---

5 At least I and the other technical reviewers didn't know it until Caleb Hattingh told us. Thanks, Caleb!

the same time. It's easier to use than the standard library's `fileinput.input` function (*https://fpy.li/18-12*) (which also provides a context manager, by the way).

If you want to study Martijn's `inplace` source code (listed in the post (*https://fpy.li/18-11*)), find the `yield` keyword: everything before it deals with setting up the context, which entails creating a backup file, then opening and yielding references to the readable and writable file handles that will be returned by the __enter__ call. The __exit__ processing after the `yield` closes the file handles and restores the file from the backup if something went wrong.

This concludes our overview of the `with` statement and context managers. Let's turn to `match/case` in the context of a complete example.

Pattern Matching in lis.py: A Case Study

In "Pattern Matching Sequences in an Interpreter" on page 43 we saw examples of sequence patterns extracted from the `evaluate` function of Peter Norvig's *lis.py* interpreter, ported to Python 3.10. In this section I want to give a broader overview of how *lis.py* works, and also explore all the `case` clauses of `evaluate`, explaining not only the patterns but also what the interpreter does in each `case`.

Besides showing more pattern matching, I wrote this section for three reasons:

1. Norvig's *lis.py* is a beautiful example of idiomatic Python code.
2. The simplicity of Scheme is a master class of language design.
3. Learning how an interpreter works gave me a deeper understanding of Python and programming languages in general—interpreted or compiled.

Before looking at the Python code, let's get a little taste of Scheme so you can make sense of this case study—in case you haven't seen Scheme or Lisp before.

Scheme Syntax

In Scheme there is no distinction between expressions and statements, like we have in Python. Also, there are no infix operators. All expressions use prefix notation like (+ x 13) instead of x + 13. The same prefix notation is used for function calls—e.g., (gcd x 13)—and special forms—e.g., (define x 13), which we'd write as the assignment statement x = 13 in Python. The notation used by Scheme and most Lisp dialects is known as *S-expression*.[6]

---

[6] People complain about too many parentheses in Lisp, but thoughtful indentation and a good editor mostly take care of that issue. The main readability problem is using the same (f …) notation for function calls and special forms like (define …), (if …), and (quote …) that don't behave at all like function calls.

Example 18-10 shows a simple example in Scheme.

Example 18-10. Greatest common divisor in Scheme

```scheme
(define (mod m n)
    (- m (* n (quotient m n))))

(define (gcd m n)
    (if (= n 0)
        m
        (gcd n (mod m n))))

(display (gcd 18 45))
```

Example 18-10 shows three Scheme expressions: two function definitions—mod and gcd—and a call to display, which will output 9, the result of (gcd 18 45). Example 18-11 is the same code in Python (shorter than an English explanation of the recursive *Euclidean algorithm (https://fpy.li/18-14)).

Example 18-11. Same as Example 18-10, written in Python

```python
def mod(m, n):
    return m - (m // n * n)

def gcd(m, n):
    if n == 0:
        return m
    else:
        return gcd(n, mod(m, n))

print(gcd(18, 45))
```

In idiomatic Python, I'd use the % operator instead of reinventing mod, and it would be more efficient to use a while loop instead of recursion. But I wanted to show two function definitions, and make the examples as similar as possible, to help you read the Scheme code.

Scheme has no iterative control flow commands like while or for. Iteration is done with recursion. Note how there are no assignments in the Scheme and Python examples. Extensive use of recursion and minimal use of assignment are hallmarks of programming in a functional style.[7]

---

7 To make iteration through recursion practical and efficient, Scheme and other functional languages implement *proper tail calls*. For more about this, see "Soapbox" on page 695.

Now let's review the code of the Python 3.10 version of *lis.py*. The complete source code with tests is in the *18-with-match/lispy/py3.10/* (*https://fpy.li/18-15*) directory of the GitHub repository *fluentpython/example-code-2e* (*https://fpy.li/code*).

Imports and Types

Example 18-12 shows the first lines of *lis.py*. The use of `TypeAlias` and the `|` type union operator require Python 3.10.

Example 18-12. lis.py: top of the file

```python
import math
import operator as op
from collections import ChainMap
from itertools import chain
from typing import Any, TypeAlias, NoReturn

Symbol: TypeAlias = str
Atom: TypeAlias = float | int | Symbol
Expression: TypeAlias = Atom | list
```

The types defined are:

Symbol

Just an alias for `str`. In *lis.py*, `Symbol` is used for identifiers; there is no string data type with operations such as slicing, splitting, etc.[8]

Atom

A simple syntactic element, such as a number or a `Symbol`—as opposed to a composite structure made of distinct parts, like a list.

Expression

The building blocks of Scheme programs are expressions made of atoms and lists, possibly nested.

The Parser

Norvig's parser is 36 lines of code showcasing the power of Python applied to handling the simple recursive syntax of S-expression—without string data, comments, macros, and other features of standard Scheme that make parsing more complicated (Example 18-13).

---

8 But Norvig's second interpreter, *lispy.py* (*https://fpy.li/18-16*), supports strings as a data type, as well as advanced features like syntactic macros, continuations, and proper tail calls. However, *lispy.py* is almost three times longer than *lis.py*—and much harder to understand.

Example 18-13. lis.py: the main parsing functions

```python
def parse(program: str) -> Expression:
    "Read a Scheme expression from a string."
    return read_from_tokens(tokenize(program))

def tokenize(s: str) -> list[str]:
    "Convert a string into a list of tokens."
    return s.replace('(', ' ( ').replace(')', ' ) ').split()

def read_from_tokens(tokens: list[str]) -> Expression:
    "Read an expression from a sequence of tokens."
    # more parsing code omitted in book listing
```

The main function of that group is `parse`, which takes an S-expression as a `str` and returns an `Expression` object, as defined in Example 18-12: an `Atom` or a `list` that may contain more atoms and nested lists.

Norvig uses a smart trick in `tokenize`: he adds spaces before and after each parenthesis in the input and then splits it, resulting in a list of syntactic tokens with `'('` and `')'` as separate tokens. This shortcut works because there is no string type in the little Scheme of *lis.py*, so every `'('` or `')'` is an expression delimiter. The recursive parsing code is in `read_from_tokens`, a 14-line function that you can read in the *fluentpython/example-code-2e* (*https://fpy.li/18-17*) repository. I will skip it because I want to focus on the other parts of the interpreter.

Here are some doctests extracted from *lispy/py3.10/examples_test.py* (*https://fpy.li/18-18*):

```python
>>> from lis import parse
>>> parse('1.5')
1.5
>>> parse('ni!')
'ni!'
>>> parse('(gcd 18 45)')
['gcd', 18, 45]
>>> parse('''
... (define double
...     (lambda (n)
...         (* n 2)))
... ''')
['define', 'double', ['lambda', ['n'], ['*', 'n', 2]]]
```

The parsing rules for this subset of Scheme are simple:

1. A token that looks like a number is parsed as a `float` or `int`.

2. Anything else that is not `'('` or `')'` is parsed as a `Symbol`—a `str` to be used as an identifier. This includes source text like +, set!, and make-counter that are valid identifiers in Scheme but not in Python.

3. Expressions inside '(' and ')' are recursively parsed as lists containing atoms or as nested lists that may contain atoms and more nested lists.

Using the terminology of the Python interpreter, the output of parse is an AST (Abstract Syntax Tree): a convenient representation of the Scheme program as nested lists forming a tree-like structure, where the outermost list is the trunk, inner lists are the branches, and atoms are the leaves (Figure 18-1).

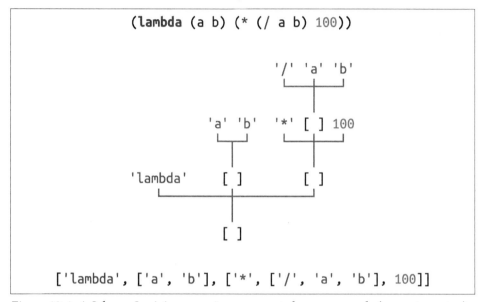

Figure 18-1. A Scheme lambda expression represented as source code (concrete syntax), as a tree, and as a sequence of Python objects (abstract syntax).

The Environment

The Environment class extends collections.ChainMap, adding a change method to update a value inside one of the chained dicts, which ChainMap instances hold in a list of mappings: the self.maps attribute. The change method is needed to support the Scheme (set! …) form, described later; see Example 18-14.

Example 18-14. lis.py: the Environment class

```python
class Environment(ChainMap[Symbol, Any]):
    "A ChainMap that allows changing an item in-place."

    def change(self, key: Symbol, value: Any) -> None:
        "Find where key is defined and change the value there."
        for map in self.maps:
            if key in map:
```

```
        map[key] = value  # type: ignore[index]
        return
    raise KeyError(key)
```

Note that the change method only updates existing keys.[9] Trying to change a key that is not found raises KeyError.

This doctest shows how Environment works:

```
>>> from lis import Environment
>>> inner_env = {'a': 2}
>>> outer_env = {'a': 0, 'b': 1}
>>> env = Environment(inner_env, outer_env)
>>> env['a']  ❶
2
>>> env['a'] = 111  ❷
>>> env['c'] = 222
>>> env
Environment({'a': 111, 'c': 222}, {'a': 0, 'b': 1})
>>> env.change('b', 333)  ❸
>>> env
Environment({'a': 111, 'c': 222}, {'a': 0, 'b': 333})
```

❶ When reading values, Environment works as ChainMap: keys are searched in the nested mappings from left to right. That's why the value of a in the outer_env is shadowed by the value in inner_env.

❷ Assigning with [] overwrites or inserts new items, but always in the first mapping, inner_env in this example.

❸ env.change('b', 333) seeks the 'b' key and assigns a new value to it in-place, in the outer_env.

Next is the standard_env() function, which builds and returns an Environment loaded with predefined functions, similar to Python's __builtins__ module that is always available (Example 18-15).

Example 18-15. lis.py: standard_env() builds and returns the global environment

```
def standard_env() -> Environment:
    "An environment with some Scheme standard procedures."
    env = Environment()
```

---

9 The # type: ignore[index] comment is there because of *typeshed* issue #6042 (*https://fpy.li/18-19*), which is unresolved as I review this chapter. ChainMap is annotated as MutableMapping, but the type hint in the maps attribute says it's a list of Mapping, indirectly making the whole ChainMap immutable as far as Mypy is concerned.

```
env.update(vars(math))   # sin, cos, sqrt, pi, ...
env.update({
        '+': op.add,
        '-': op.sub,
        '*': op.mul,
        '/': op.truediv,
        # omitted here: more operator definitions
        'abs': abs,
        'append': lambda *args: list(chain(*args)),
        'apply': lambda proc, args: proc(*args),
        'begin': lambda *x: x[-1],
        'car': lambda x: x[0],
        'cdr': lambda x: x[1:],
        # omitted here: more function definitions
        'number?': lambda x: isinstance(x, (int, float)),
        'procedure?': callable,
        'round': round,
        'symbol?': lambda x: isinstance(x, Symbol),
})
return env
```

To summarize, the env mapping is loaded with:

- All functions from Python's math module

- Selected operators from Python's op module

- Simple but powerful functions built with Python's lambda

- Python built-ins renamed, like callable as procedure?, or directly mapped, like round

The REPL

Norvig's REPL (read-eval-print-loop) is easy to understand but not user-friendly (see Example 18-16). If no command-line arguments are given to *lis.py*, the repl() function is invoked by main()—defined at the end of the module. At the lis.py> prompt, we must enter correct and complete expressions; if we forget to close one parenthesis, *lis.py* crashes.[10]

---

10 As I studied Norvig's *lis.py* and *lispy.py*, I started a fork named *mylis* (*https://fpy.li/18-20*) that adds some features, including a REPL that accepts partial S-expressions and prompts for the continuation, similar to how Python's REPL knows we are not finished and presents the secondary prompt (...) until we enter a complete expression or statement that can be evaluated. *mylis* also handles a few errors gracefully, but it's still easy to crash. It's not nearly as robust as Python's REPL.

Example 18-16. The REPL functions

```python
def repl(prompt: str = 'lis.py> ') -> NoReturn:
    "A prompt-read-eval-print loop."
    global_env = Environment({}, standard_env())
    while True:
        ast = parse(input(prompt))
        val = evaluate(ast, global_env)
        if val is not None:
            print(lispstr(val))

def lispstr(exp: object) -> str:
    "Convert a Python object back into a Lisp-readable string."
    if isinstance(exp, list):
        return '(' + ' '.join(map(lispstr, exp)) + ')'
    else:
        return str(exp)
```

Here is a quick explanation about these two functions:

`repl(prompt: str = 'lis.py> ') -> NoReturn`
Calls `standard_env()` to provide built-in functions for the global environment, then enters an infinite loop, reading and parsing each input line, evaluating it in the global environment, and displaying the result—unless it's `None`. The `global_env` may be modified by `evaluate`. For example, when a user defines a new global variable or named function, it is stored in the first mapping of the environment—the empty `dict` in the `Environment` constructor call in the first line of `repl`.

`lispstr(exp: object) -> str`
The inverse function of `parse`: given a Python object representing an expression, `parse` returns the Scheme source code for it. For example, given `['+', 2, 3]`, the result is `'(+ 2 3)'`.

The Evaluator

Now we can appreciate the beauty of Norvig's expression evaluator—made a little prettier with `match/case`. The `evaluate` function in Example 18-17 takes an `Expres sion` built by `parse` and an `Environment`.

The body of `evaluate` is a single `match` statement with an expression `exp` as the subject. The `case` patterns express the syntax and semantics of Scheme with amazing clarity.

Example 18-17. evaluate takes an expression and computes its value

```
KEYWORDS = ['quote', 'if', 'lambda', 'define', 'set!']

def evaluate(exp: Expression, env: Environment) -> Any:
    "Evaluate an expression in an environment."
    match exp:
        case int(x) | float(x):
            return x
        case Symbol(var):
            return env[var]
        case ['quote', x]:
            return x
        case ['if', test, consequence, alternative]:
            if evaluate(test, env):
                return evaluate(consequence, env)
            else:
                return evaluate(alternative, env)
        case ['lambda', [*parms], *body] if body:
            return Procedure(parms, body, env)
        case ['define', Symbol(name), value_exp]:
            env[name] = evaluate(value_exp, env)
        case ['define', [Symbol(name), *parms], *body] if body:
            env[name] = Procedure(parms, body, env)
        case ['set!', Symbol(name), value_exp]:
            env.change(name, evaluate(value_exp, env))
        case [func_exp, *args] if func_exp not in KEYWORDS:
            proc = evaluate(func_exp, env)
            values = [evaluate(arg, env) for arg in args]
            return proc(*values)
        case _:
            raise SyntaxError(lispstr(exp))
```

Let's study each `case` clause and what it does. In some cases I added comments showing an S-expression that would match the pattern when parsed into a Python list. Doctests extracted from *examples_test.py* (*https://fpy.li/18-21*) demonstrate each case.

Evaluating numbers

```
        case int(x) | float(x):
            return x
```

Subject:
    Instance of `int` or `float`.

Action:
    Return value as is.

Example:

```
>>> from lis import parse, evaluate, standard_env
>>> evaluate(parse('1.5'), {})
1.5
```

Evaluating symbols

```
case Symbol(var):
    return env[var]
```

Subject:

Instance of Symbol, i.e., a str used as an identifier.

Action:

Look up var in env and return its value.

Examples:

```
>>> evaluate(parse('+'), standard_env())
<built-in function add>
>>> evaluate(parse('ni!'), standard_env())
Traceback (most recent call last):
    ...
KeyError: 'ni!'
```

(quote …)

The quote special form treats atoms and lists as data instead of expressions to be evaluated.

```
# (quote (99 bottles of beer))
case ['quote', x]:
    return x
```

Subject:

List starting with the symbol 'quote', followed by one expression x.

Action:

Return x without evaluating it.

Examples:

```
>>> evaluate(parse('(quote no-such-name)'), standard_env())
'no-such-name'
>>> evaluate(parse('(quote (99 bottles of beer))'), standard_env())
[99, 'bottles', 'of', 'beer']
>>> evaluate(parse('(quote (/ 10 0))'), standard_env())
['/', 10, 0]
```

Without quote, each expression in the test would raise an error:

- `no-such-name` would be looked up in the environment, raising `KeyError`
- `(99 bottles of beer)` cannot be evaluated because the number 99 is not a `Symbol` naming a special form, operator, or function
- `(/ 10 0)` would raise `ZeroDivisionError`

Why Languages Have Reserved Keywords

Although simple, `quote` cannot be implemented as a function in Scheme. Its special power is to prevent the interpreter from evaluating `(f 10)` in the expression `(quote (f 10))`: the result is simply a list with a `Symbol` and an `int`. In contrast, in a function call like `(abs (f 10))`, the interpreter evaluates `(f 10)` before invoking `abs`. That's why `quote` is a reserved keyword: it must be handled as a special form.

In general, reserved keywords are needed:

- To introduce specialized evaluation rules, as in `quote` and `lambda`—which don't evaluate any of their subexpressions
- To change the control flow, as in `if` and function calls—which also have special evaluation rules
- To manage the environment, as in `define` and `set`

This is also why Python, and programming languages in general, need reserved keywords. Think about Python's `def`, `if`, `yield`, `import`, `del`, and what they do.

(if ...)

```
# (if (< x 0) 0 x)
case ['if', test, consequence, alternative]:
    if evaluate(test, env):
        return evaluate(consequence, env)
    else:
        return evaluate(alternative, env)
```

Subject:

List starting with `'if'` followed by three expressions: `test`, `consequence`, and `alternative`.

Action:

Evaluate `test`:

- If true, evaluate `consequence` and return its value.
- Otherwise, evaluate `alternative` and return its value.

Examples:

```
>>> evaluate(parse('(if (= 3 3) 1 0))'), standard_env())
1
>>> evaluate(parse('(if (= 3 4) 1 0))'), standard_env())
0
```

The `consequence` and `alternative` branches must be single expressions. If more than one expression is needed in a branch, you can combine them with (`begin exp1 exp2…`), provided as a function in *lis.py*—see Example 18-15.

(lambda ...)

Scheme's `lambda` form defines anonymous functions. It doesn't suffer from the limitations of Python's `lambda`: any function that can be written in Scheme can be written using the (`lambda …`) syntax.

```
# (lambda (a b) (/ (+ a b) 2))
case ['lambda' [*parms], *body] if body:
    return Procedure(parms, body, env)
```

Subject:

List starting with `'lambda'`, followed by:

- List of zero or more parameter names.

- One or more expressions collected in `body` (the guard ensures that `body` is not empty).

Action:

Create and return a new `Procedure` instance with the parameter names, the list of expressions as the body, and the current environment.

Example:

```
>>> expr = '(lambda (a b) (* (/ a b) 100))'
>>> f = evaluate(parse(expr), standard_env())
>>> f   # doctest: +ELLIPSIS
<lis.Procedure object at 0x...>
>>> f(15, 20)
75.0
```

The `Procedure` class implements the concept of a closure: a callable object holding parameter names, a function body, and a reference to the environment in which the function is defined. We'll study the code for `Procedure` in a moment.

(define ...)

The define keyword is used in two different syntactic forms. The simplest is:

```
# (define half (/ 1 2))
case ['define', Symbol(name), value_exp]:
    env[name] = evaluate(value_exp, env)
```

Subject:

List starting with 'define', followed by a Symbol and an expression.

Action:

Evaluate the expression and put its value into env, using name as key.

Example:

```
>>> global_env = standard_env()
>>> evaluate(parse('(define answer (* 7 6))'), global_env)
>>> global_env['answer']
42
```

The doctest for this case creates a global_env so that we can verify that evaluate puts answer into that Environment.

We can use that simple define form to create variables or to bind names to anonymous functions, using (lambda ...) as the value_exp.

Standard Scheme provides a shortcut for defining named functions. That's the second define form:

```
# (define (average a b) (/ (+ a b) 2))
case ['define', [Symbol(name), *parms], *body] if body:
    env[name] = Procedure(parms, body, env)
```

Subject:

List starting with 'define', followed by:

- A list starting with a Symbol(name), followed by zero or more items collected into a list named parms.

- One or more expressions collected in body (the guard ensures that body is not empty).

Action:

- Create a new Procedure instance with the parameter names, the list of expressions as the body, and the current environment.

- Put the Procedure into env, using name as key.

The doctest in Example 18-18 defines a function named % that computes a percentage and adds it to the global_env.

Example 18-18. Defining a function named % that computes a percentage

```
>>> global_env = standard_env()
>>> percent = '(define (% a b) (* (/ a b) 100))'
>>> evaluate(parse(percent), global_env)
>>> global_env['%']  # doctest: +ELLIPSIS
<lis.Procedure object at 0x...>
>>> global_env['%'](170, 200)
85.0
```

After calling `evaluate`, we check that `%` is bound to a `Procedure` that takes two numeric arguments and returns a percentage.

The pattern for the second `define` case does not enforce that the items in `parms` are all `Symbol` instances. I'd have to check that before building the `Procedure`, but I didn't—to keep the code as easy to follow as Norvig's.

(set! …)

The `set!` form changes the value of a previously defined variable.[11]

```
# (set! n (+ n 1))
case ['set!', Symbol(name), value_exp]:
    env.change(name, evaluate(value_exp, env))
```

Subject:

List starting with `'set!'`, followed by a `Symbol` and an expression.

Action:

Update the value of `name` in `env` with the result of evaluating the expression.

The `Environment.change` method traverses the chained environments from local to global, and updates the first occurrence of `name` with the new value. If we were not implementing the `'set!'` keyword, we could use Python's `ChainMap` as the `Environment` type everywhere in this interpreter.

---

11 Assignment is one of the first features taught in many programming tutorials, but `set!` only appears on page 220 of the best known Scheme book, *Structure and Interpretation of Computer Programs*, 2nd ed., (*https://fpy.li/18-22*) by Abelson et al. (MIT Press), a.k.a. SICP or the "Wizard Book." Coding in a functional style can take us very far without the state changes that are typical of imperative and object-oriented programming.

Python's nonlocal and Scheme's set! Address the Same Issue

The use of the set! form is related to the use of the nonlocal keyword in Python: declaring nonlocal x allows x = 10 to update a previously defined x variable outside of the local scope. Without a nonlocal x declaration, x = 10 will always create a local variable in Python, as we saw in "The nonlocal Declaration" on page 317.

Similarly, (set! x 10) updates a previously defined x that may be outside of the local environment of the function. In contrast, the variable x in (define x 10) is always a local variable, created or updated in the local environment.

Both nonlocal and (set! …) are needed to update program state held in variables within a closure. Example 9-13 demonstrated the use of nonlocal to implement a function to compute a running average, holding an item count and total in a closure. Here is that same idea, written in the Scheme subset of *lis.py*:

```
(define (make-averager)
    (define count 0)
    (define total 0)
    (lambda (new-value)
        (set! count (+ count 1))
        (set! total (+ total new-value))
        (/ total count)
    )
)
(define avg (make-averager))  ❶
(avg 10)  ❷
(avg 11)  ❸
(avg 15)  ❹
```

❶ Creates a new closure with the inner function defined by lambda, and the variables count and total initialized to 0; binds the closure to avg.

❷ Returns 10.0.

❸ Returns 10.5.

❹ Returns 12.0.

The preceding code is one of the tests in *lispy/py3.10/examples_test.py* (*https://fpy.li/ 18-18*).

Now we get to a function call.

Function call

```
# (gcd (* 2 105) 84)
case [func_exp, *args] if func_exp not in KEYWORDS:
    proc = evaluate(func_exp, env)
    values = [evaluate(arg, env) for arg in args]
    return proc(*values)
```

Subject:

List with one or more items.

The guard ensures that func_exp is not one of ['quote', 'if', 'define', 'lambda', 'set!']—listed right before evaluate in Example 18-17.

The pattern matches any list with one or more expressions, binding the first expression to func_exp and the rest to args as a list, which may be empty.

Action:

- Evaluate func_exp to obtain a function proc.

- Evaluate each item in args to build a list of argument values.

- Call proc with the values as separate arguments, returning the result.

Example:

```
>>> evaluate(parse('(% (* 12 14) (- 500 100))'), global_env)
42.0
```

This doctest continues from Example 18-18: it assumes global_env has a function named %. The arguments given to % are arithmetic expressions, to emphasize that the arguments are evaluated before the function is called.

The guard in this case is needed because [func_exp, *args] matches any sequence subject with one or more items. However, if func_exp is a keyword, and the subject did not match any previous case, then it is really a syntax error.

Catch syntax errors

If the subject exp does not match any of the previous cases, the catch-all case raises a SyntaxError:

```
case _:
    raise SyntaxError(lispstr(exp))
```

Here is an example of a malformed (lambda …) reported as a SyntaxError:

```
>>> evaluate(parse('(lambda is not like this)'), standard_env())
Traceback (most recent call last):
  ...
SyntaxError: (lambda is not like this)
```

If the case for function call did not have that guard rejecting keywords, the (lambda is not like this) expression would be handled as a function call, which would raise KeyError because 'lambda' is not part of the environment—just like lambda is not a Python built-in function.

Procedure: A Class Implementing a Closure

The Procedure class could very well be named Closure, because that's what it represents: a function definition together with an environment. The function definition includes the name of the parameters and the expressions that make up the body of the function. The environment is used when the function is called to provide the values of the *free variables*: variables that appear in the body of the function but are not parameters, local variables, or global variables. We saw the concepts of *closure* and *free variable* in "Closures" on page 313.

We learned how to use closures in Python, but now we can dive deeper and see how a closure is implemented in *lis.py*:

```
class Procedure:
    "A user-defined Scheme procedure."

    def __init__(   ❶
        self, parms: list[Symbol], body: list[Expression], env: Environment
    ):
        self.parms = parms   ❷
        self.body = body
        self.env = env

    def __call__(self, *args: Expression) -> Any:   ❸
        local_env = dict(zip(self.parms, args))   ❹
        env = Environment(local_env, self.env)   ❺
        for exp in self.body:   ❻
            result = evaluate(exp, env)
        return result   ❼
```

❶ Called when a function is defined by the lambda or define forms.

❷ Save the parameter names, body expressions, and environment for later use.

❸ Called by proc(*values) in the last line of the case [func_exp, *args] clause.

❹ Build local_env mapping self.parms as local variable names, and the given args as values.

❺ Build a new combined env, putting local_env first, and then self.env—the environment that was saved when the function was defined.

❻ Iterate over each expression in self.body, evaluating it in the combined env.

❼ Return the result of the last expression evaluated.

There are a couple of simple functions after evaluate in *lis.py* (*https://fpy.li/18-24*): run reads a complete Scheme program and executes it, and main calls run or repl, depending on the command line—similar to what Python does. I will not describe those functions because there's nothing new in them. My goals were to share with you the beauty of Norvig's little interpreter, to give more insight into how closures work, and to show how match/case is a great addition to Python.

To wrap up this extended section on pattern matching, let's formalize the concept of an OR-pattern.

Using OR-patterns

A series of patterns separated by | is an *OR-pattern* (*https://fpy.li/18-25*): it succeeds if any of the subpatterns succeed. The pattern in "Evaluating numbers" on page 681 is an OR-pattern:

```
case int(x) | float(x):
    return x
```

All subpatterns in an OR-pattern must use the same variables. This restriction is necessary to ensure that the variables are available to the guard expression and the case body, regardless of the subpattern that matched.

In the context of a case clause, the | operator has a special meaning. It does not trigger the __or__ special method, which handles expressions like a | b in other contexts, where it is overloaded to perform operations such as set union or integer bitwise-or, depending on the operands.

An OR-pattern is not restricted to appear at the top level of a pattern. You can also use | in subpatterns. For example, if we wanted *lis.py* to accept the Greek letter λ (lambda)[12] as well as the lambda keyword, we can rewrite the pattern like this:

```
# (λ (a b) (/ (+ a b) 2) )
case ['lambda' | 'λ', [*parms], *body] if body:
    return Procedure(parms, body, env)
```

---

12 The official Unicode name for λ (U+03BB) is GREEK SMALL LETTER LAMDA. This is not a typo: the character is named "lamda" without the "b" in the Unicode database. According to the English Wikipedia article "Lambda" (*https://fpy.li/18-26*), the Unicode Consortium adopted that spelling because of "preferences expressed by the Greek National Body."

Now we can move to the third and last subject of this chapter: the unusual places where an else clause may appear in Python.

Do This, Then That: else Blocks Beyond if

This is no secret, but it is an underappreciated language feature: the else clause can be used not only in if statements but also in for, while, and try statements.

The semantics of for/else, while/else, and try/else are closely related, but very different from if/else. Initially, the word else actually hindered my understanding of these features, but eventually I got used to it.

Here are the rules:

for
> The else block will run only if and when the for loop runs to completion (i.e., not if the for is aborted with a break).

while
> The else block will run only if and when the while loop exits because the condition became *falsy* (i.e., not if the while is aborted with a break).

try
> The else block will run only if no exception is raised in the try block. The official docs (*https://fpy.li/18-27*) also state: "Exceptions in the else clause are not handled by the preceding except clauses."

In all cases, the else clause is also skipped if an exception or a return, break, or continue statement causes control to jump out of the main block of the compound statement.

 I think else is a very poor choice for the keyword in all cases except if. It implies an excluding alternative, like, "Run this loop, otherwise do that," but the semantics for else in loops is the opposite: "Run this loop, then do that." This suggests then as a better keyword—which would also make sense in the try context: "Try this, then do that." However, adding a new keyword is a breaking change to the language—not an easy decision to make.

Using else with these statements often makes the code easier to read and saves the trouble of setting up control flags or coding extra if statements.

The use of else in loops generally follows the pattern of this snippet:

```
for item in my_list:
    if item.flavor == 'banana':
```

```
        break
    else:
        raise ValueError('No banana flavor found!')
```

In the case of try/except blocks, else may seem redundant at first. After all, the after_call() in the following snippet will run only if the dangerous_call() does not raise an exception, correct?

```
try:
    dangerous_call()
    after_call()
except OSError:
    log('OSError...')
```

However, doing so puts the after_call() inside the try block for no good reason. For clarity and correctness, the body of a try block should only have the statements that may generate the expected exceptions. This is better:

```
try:
    dangerous_call()
except OSError:
    log('OSError...')
else:
    after_call()
```

Now it's clear that the try block is guarding against possible errors in dangerous_call() and not in after_call(). It's also explicit that after_call() will only execute if no exceptions are raised in the try block.

In Python, try/except is commonly used for control flow, and not just for error handling. There's even an acronym/slogan for that documented in the official Python glossary (*https://fpy.li/18-28*):

EAFP

> Easier to ask for forgiveness than permission. This common Python coding style assumes the existence of valid keys or attributes and catches exceptions if the assumption proves false. This clean and fast style is characterized by the presence of many try and except statements. The technique contrasts with the *LBYL* style common to many other languages such as C.

The glossary then defines LBYL:

LBYL

> Look before you leap. This coding style explicitly tests for pre-conditions before making calls or lookups. This style contrasts with the *EAFP* approach and is characterized by the presence of many if statements. In a multi-threaded environment, the LBYL approach can risk introducing a race condition between "the looking" and "the leaping." For example, the code, if key in mapping: return mapping[key] can fail if another thread removes key from mapping after the test, but before the lookup. This issue can be solved with locks or by using the EAFP approach.

Given the EAFP style, it makes even more sense to know and use else blocks well in try/except statements.

 When the match statement was discussed, some people (including me) thought it should also have an else clause. In the end it was decided that it wasn't needed because case _: does the same job.[13]

Now let's summarize the chapter.

Chapter Summary

This chapter started with context managers and the meaning of the with statement, quickly moving beyond its common use to automatically close opened files. We implemented a custom context manager: the LookingGlass class with the __enter__/__exit__ methods, and saw how to handle exceptions in the __exit__ method. A key point that Raymond Hettinger made in his PyCon US 2013 keynote is that with is not just for resource management; it's a tool for factoring out common setup and teardown code, or any pair of operations that need to be done before and after another procedure.[14]

We reviewed functions in the contextlib standard library module. One of them, the @contextmanager decorator, makes it possible to implement a context manager using a simple generator with one yield—a leaner solution than coding a class with at least two methods. We reimplemented the LookingGlass as a looking_glass generator function, and discussed how to do exception handling when using @contextmanager.

Then we studied Peter Norvig's elegant *lis.py*, a Scheme interpreter written in idiomatic Python, refactored to use match/case in evaluate—the function at the core of any interpreter. Understanding how evaluate works required reviewing a little bit of Scheme, a parser for S-expressions, a simple REPL, and the construction of nested scopes through an Environment subclass of collection.ChainMap. In the end, *lis.py* became a vehicle to explore much more than pattern matching. It shows how the different parts of an interpreter work together, illuminating core features of Python itself: why reserved keywords are necessary, how scoping rules work, and how closures are built and used.

---

13 Watching the discussion in the python-dev mailing list I thought one reason why else was rejected was the lack of consensus on how to indent it within match: should else be indented at the same level as match, or at the same level as case?

14 See slide 21 in "Python is Awesome" (*https://fpy.li/18-29*).

Further Reading

Chapter 8, "Compound Statements," (*https://fpy.li/18-27*) in *The Python Language Reference* says pretty much everything there is to say about else clauses in if, for, while, and try statements. Regarding Pythonic usage of try/except, with or without else, Raymond Hettinger has a brilliant answer to the question "Is it a good practice to use try-except-else in Python?" (*https://fpy.li/18-31*) in StackOverflow. *Python in a Nutshell*, 3rd ed., by Martelli et al., has a chapter about exceptions with an excellent discussion of the EAFP style, crediting computing pioneer Grace Hopper for coining the phrase, "It's easier to ask forgiveness than permission."

The Python Standard Library, Chapter 4, "Built-in Types," has a section devoted to "Context Manager Types" (*https://fpy.li/18-32*). The __enter__/__exit__ special methods are also documented in *The Python Language Reference* in "With Statement Context Managers" (*https://fpy.li/18-33*). Context managers were introduced in PEP 343—The "with" Statement (*https://fpy.li/pep343*).

Raymond Hettinger highlighted the with statement as a "winning language feature" in his PyCon US 2013 keynote (*https://fpy.li/18-29*). He also showed some interesting applications of context managers in his talk, "Transforming Code into Beautiful, Idiomatic Python" (*https://fpy.li/18-35*), at the same conference.

Jeff Preshing's blog post "The Python *with* Statement by Example" (*https://fpy.li/18-36*) is interesting for the examples using context managers with the pycairo graphics library.

The contextlib.ExitStack class is based on an original idea by Nikolaus Rath, who wrote a short post explaining why its useful: "On the Beauty of Python's ExitStack" (*https://fpy.li/18-37*). In that text, Rath submits that ExitStack is similar but more flexible than the defer statement in Go—which I think is one of the best ideas in that language.

Beazley and Jones devised context managers for very different purposes in their *Python Cookbook,* 3rd ed. "Recipe 8.3. Making Objects Support the Context-Management Protocol" implements a LazyConnection class whose instances are context managers that open and close network connections automatically in with blocks. "Recipe 9.22. Defining Context Managers the Easy Way" introduces a context manager for timing code, and another for making transactional changes to a list object: within the with block, a working copy of the list instance is made, and all changes are applied to that working copy. Only when the with block completes without an exception, the working copy replaces the original list. Simple and ingenious.

Peter Norvig describes his small Scheme interpreters in the posts "(How to Write a (Lisp) Interpreter (in Python))" (*https://fpy.li/18-38*) and "(An ((Even Better) Lisp) Interpreter (in Python))" (*https://fpy.li/18-39*). The code for *lis.py* and *lispy.py* is the *norvig/pytudes* (*https://fpy.li/18-40*) repository. My repository *fluentpython/lispy* (*https://fpy.li/18-41*) includes the *mylis* forks of *lis.py*, updated to Python 3.10, with a nicer REPL, command-line integration, examples, more tests, and references for learning more about Scheme. The best Scheme dialect and environment to learn and experiment is Racket (*https://fpy.li/18-42*).

Soapbox

Factoring Out the Bread

In his PyCon US 2013 keynote, "What Makes Python Awesome" (*https://fpy.li/18-1*), Raymond Hettinger says when he first saw the with statement proposal he thought it was "a little bit arcane." Initially, I had a similar reaction. PEPs are often hard to read, and PEP 343 is typical in that regard.

Then—Hettinger told us—he had an insight: subroutines are the most important invention in the history of computer languages. If you have sequences of operations like A;B;C and P;B;Q, you can factor out B in a subroutine. It's like factoring out the filling in a sandwich: using tuna with different breads. But what if you want to factor out the bread, to make sandwiches with wheat bread, using a different filling each time? That's what the with statement offers. It's the complement of the subroutine. Hettinger went on to say:

> The with statement is a very big deal. I encourage you to go out and take this tip of the iceberg and drill deeper. You can probably do profound things with the with statement. The best uses of it have not been discovered yet. I expect that if you make good use of it, it will be copied into other languages and all future languages will have it. You can be part of discovering something almost as profound as the invention of the subroutine itself.

Hettinger admits he is overselling the with statement. Nevertheless, it is a very useful feature. When he used the sandwich analogy to explain how with is the complement to the subroutine, many possibilities opened up in my mind.

If you need to convince anyone that Python is awesome, you should watch Hettinger's keynote. The bit about context managers is from 23:00 to 26:15. But the entire keynote is excellent.

Efficient Recursion with Proper Tail Calls

Standard Scheme implementations are required to provide *proper tail calls* (PTC), to make iteration through recursion a practical alternative to while loops in imperative languages. Some writers refer to PTC as *tail call optimization* (TCO); for others, TCO is something different. For more details, see "Tail call" (*https://fpy.li/18-44*) on Wikipedia and "Tail call optimization in ECMAScript 6" (*https://fpy.li/18-45*).

A *tail call* is when a function returns the result of a function call, which may be the same function or not. The gcd examples in Example 18-10 and Example 18-11 make (recursive) tail calls in the *falsy* branch of the if.

On the other hand, this factorial does not make a tail call:

```
def factorial(n):
    if n < 2:
        return 1
    return n * factorial(n - 1)
```

The call to factorial in the last line is not a tail call because the return value is not the result of the recursive call: the result is multiplied by n before it is returned.

Here is an alternative that uses a tail call, and is therefore *tail recursive*:

```
def factorial_tc(n, product=1):
    if n < 1:
        return product
    return factorial_tc(n - 1, product * n)
```

Python does not have PTC, so there's no advantage in writing tail recursive functions. In this case, the first version is shorter and more readable in my opinion. For real-life uses, don't forget that Python has math.factorial, written in C without recursion. The point is that, even in languages that implement PTC, it does not benefit every recursive function, only those that are carefully written to make tail calls.

If PTC is supported by the language, when the interpreter sees a tail call, it jumps into the body of the called function without creating a new stack frame, saving memory. There are also compiled languages that implement PTC, sometimes as an optimization that can be toggled.

There is no universal consensus about the definition of TCO or the value of PTC in languages that were not designed as functional languages from the ground up, like Python or JavaScript. In functional languages, PTC is an expected feature, not merely an optimization that is nice to have. If a language has no iteration mechanism other than recursion, then PTC is necessary for practical usage. Norvig's *lis.py* (*https://fpy.li/18-46*) does not implement PTC, but his more elaborate *lispy.py* (*https://fpy.li/18-16*) interpreter does.

The Case Against Proper Tail Calls in Python and JavaScript

CPython does not implement PTC, and probably never will. Guido van Rossum wrote "Final Words on Tail Calls" (*https://fpy.li/18-48*) to explain why. To summarize, here is a key passage from his post:

> Personally, I think it is a fine feature for some languages, but I don't think it fits Python: the elimination of stack traces for some calls but not others would certainly confuse many users, who have not been raised with tail call religion but might have learned about call semantics by tracing through a few calls in a debugger.

In 2015, PTC was included in the ECMAScript 6 standard for JavaScript. As of October 2021, the interpreter in WebKit implements it (*https://fpy.li/18-49*). WebKit is used by Safari. The JS interpreters in every other major browser don't have PTC, and neither does Node.js, as it relies on the V8 engine that Google maintains for Chrome. Transpilers and polyfills targeting JS, like TypeScript, ClojureScript, and Babel, don't support PTC either, according to this "ECMAScript 6 compatibility table" (*https://fpy.li/18-50*).

I've seen several explanations for the rejection of PTC by the implementers, but the most common is the same that Guido van Rossum mentioned: PTC makes debugging harder for everyone, while benefiting only a minority of people who'd rather use recursion for iteration. For details, see "What happened to proper tail calls in JavaScript?" (*https://fpy.li/18-51*) by Graham Marlow.

There are cases when recursion is the best solution, even in Python without PTC. In a previous post (*https://fpy.li/18-52*) on the subject, Guido wrote:

> [...] a typical Python implementation allows 1000 recursions, which is plenty for non-recursively written code and for code that recourses to traverse, for example, a typical parse tree, but not enough for a recursively written loop over a large list.

I agree with Guido and the majority of JS implementers: PTC is not a good fit for Python or JavaScript. The lack of PTC is the main restriction for writing Python programs in a functional style—more than the limited `lambda` syntax.

If you are curious to see how PTC works in an interpreter with less features (and less code) than Norvig's *lispy.py*, check out *mylis_2* (*https://fpy.li/18-53*). The trick starts with the infinite loop in `evaluate` and the code in the `case` for function calls: that combination makes the intepreter jump into the body of the next `Procedure` without calling `evaluate` recursively during a tail call. Those little interpreters demonstrate the power of abstraction: even though Python does not implement PTC, it's possible and not very hard to write an interpreter, in Python, that does implement PTC. I learned how to do it reading Peter Norvig's code. Thanks for sharing it, professor!

Norvig's Take on evaluate() with Pattern Matching

I shared the code for the Python 3.10 version of *lis.py* with Peter Norvig. He liked the example using pattern matching, but suggested a different solution: instead of the guards I wrote, he would have exactly one `case` per keyword, and have tests within each `case`, to provide more specific `SyntaxError` messages—for example, when a body is empty. This would also make the guard in `case [func_exp, *args]` if `func_exp not in KEYWORDS:` unnecessary, as every keyword would be handled before the `case` for function calls.

I'll probably follow Norvig's advice when I add more functionality to *mylis* (*https://fpy.li/18-54*). But the way I structured `evaluate` in Example 18-17 has some didactic advantages for this book: the example parallels the implementation with `if/elif/...`

(Example 2-11), the case clauses demonstrate more features of pattern matching, and the code is more concise.

## CHAPTER 19

Concurrency Models in Python

Concurrency is about dealing with lots of things at once.

Parallelism is about doing lots of things at once.

Not the same, but related.

One is about structure, one is about execution.

Concurrency provides a way to structure a solution to solve a problem that may (but not necessarily) be parallelizable.

— Rob Pike, co-inventor of the Go language[1]

This chapter is about how to make Python deal with "lots of things at once." This may involve concurrent or parallel programming—even academics who are keen on jargon disagree on how to use those terms. I will adopt Rob Pike's informal definitions in this chapter's epigraph, but note that I've found papers and books that claim to be about parallel computing but are mostly about concurrency.[2]

Parallelism is a special case of concurrency, in Pike's view. All parallel systems are concurrent, but not all concurrent systems are parallel. In the early 2000s we used single-core machines that handled 100 processes concurrently on GNU Linux. A modern laptop with 4 CPU cores is routinely running more than 200 processes at any given time under normal, casual use. To execute 200 tasks in parallel, you'd need 200 cores. So, in practice, most computing is concurrent and not parallel. The OS

---

1 Slide 8 of the talk "Concurrency Is Not Parallelism" (*https://fpy.li/19-1*).

2 I studied and worked with Prof. Imre Simon, who liked to say there are two major sins in science: using different words to mean the same thing and using one word to mean different things. Imre Simon (1943–2009) was a pioneer of computer science in Brazil who made seminal contributions to Automata Theory and started the field of Tropical Mathematics. He was also an advocate of free software and free culture.

manages hundreds of processes, making sure each has an opportunity to make progress, even if the CPU itself can't do more than four things at once.

This chapter assumes no prior knowledge of concurrent or parallel programming. After a brief conceptual introduction, we will study simple examples to introduce and compare Python's core packages for concurrent programming: threading, multi processing, and asyncio.

The last 30% of the chapter is a high-level overview of third-party tools, libraries, application servers, and distributed task queues—all of which can enhance the performance and scalability of Python applications. These are all important topics, but beyond the scope of a book focused on core Python language features. Nevertheless, I felt it was important to address these themes in this second edition of *Fluent Python*, because Python's fitness for concurrent and parallel computing is not limited to what the standard library provides. That's why YouTube, DropBox, Instagram, Reddit, and others were able to achieve web scale when they started, using Python as their primary language—despite persistent claims that "Python doesn't scale."

What's New in This Chapter

This chapter is new in the second edition of *Fluent Python*. The spinner examples in "A Concurrent Hello World" on page 705 previously were in the chapter about *asyncio*. Here they are improved, and provide the first illustration of Python's three approaches to concurrency: threads, processes, and native coroutines.

The remaining content is new, except for a few paragraphs that originally appeared in the chapters on concurrent.futures and *asyncio*.

"Python in the Multicore World" on page 729 is different from the rest of the book: there are no code examples. The goal is to mention important tools that you may want to study to achieve high-performance concurrency and parallelism beyond what's possible with Python's standard library.

The Big Picture

There are many factors that make concurrent programming hard, but I want to touch on the most basic factor: starting threads or processes is easy enough, but how do you keep track of them?[3]

When you call a function, the calling code is blocked until the function returns. So you know when the function is done, and you can easily get the value it returned. If

---

[3] This section was suggested by my friend Bruce Eckel—author of books about Kotlin, Scala, Java, and C++.

the function raises an exception, the calling code can surround the call site with `try/except` to catch the error.

Those familiar options are not available when you start a thread or process: you don't automatically know when it's done, and getting back results or errors requires setting up some communication channel, such as a message queue.

Additionally, starting a thread or a process is not cheap, so you don't want to start one of them just to perform a single computation and quit. Often you want to amortize the startup cost by making each thread or process into a "worker" that enters a loop and stands by for inputs to work on. This further complicates communications and introduces more questions. How do you make a worker quit when you don't need it anymore? And how do you make it quit without interrupting a job partway, leaving half-baked data and unreleased resources—like open files? Again the usual answers involve messages and queues.

A coroutine is cheap to start. If you start a coroutine using the `await` keyword, it's easy to get a value returned by it, it can be safely cancelled, and you have a clear site to catch exceptions. But coroutines are often started by the asynchronous framework, and that can make them as hard to monitor as threads or processes.

Finally, Python coroutines and threads are not suitable for CPU-intensive tasks, as we'll see.

That's why concurrent programming requires learning new concepts and coding patterns. Let's first make sure we are on the same page regarding some core concepts.

A Bit of Jargon

Here are some terms I will use for the rest of this chapter and the next two:

Concurrency
> The ability to handle multiple pending tasks, making progress one at a time or in parallel (if possible) so that each of them eventually succeeds or fails. A single-core CPU is capable of concurrency if it runs an OS scheduler that interleaves the execution of the pending tasks. Also known as multitasking.

Parallelism
> The ability to execute multiple computations at the same time. This requires a multicore CPU, multiple CPUs, a GPU (*https://fpy.li/19-2*), or multiple computers in a cluster.

Execution unit
> General term for objects that execute code concurrently, each with independent state and call stack. Python natively supports three kinds of execution units: *processes*, *threads*, and *coroutines*.

Process

An instance of a computer program while it is running, using memory and a slice of the CPU time. Modern desktop operating systems routinely manage hundreds of processes concurrently, with each process isolated in its own private memory space. Processes communicate via pipes, sockets, or memory mapped files—all of which can only carry raw bytes. Python objects must be serialized (converted) into raw bytes to pass from one process to another. This is costly, and not all Python objects are serializable. A process can spawn subprocesses, each called a child process. These are also isolated from each other and from the parent. Processes allow *preemptive multitasking*: the OS scheduler *preempts*—i.e., suspends —each running process periodically to allow other processes to run. This means that a frozen process can't freeze the whole system—in theory.

Thread

An execution unit within a single process. When a process starts, it uses a single thread: the main thread. A process can create more threads to operate concurrently by calling operating system APIs. Threads within a process share the same memory space, which holds live Python objects. This allows easy data sharing between threads, but can also lead to corrupted data when more than one thread updates the same object concurrently. Like processes, threads also enable *preemptive multitasking* under the supervision of the OS scheduler. A thread consumes less resources than a process doing the same job.

Coroutine

A function that can suspend itself and resume later. In Python, *classic coroutines* are built from generator functions, and *native coroutines* are defined with `async def`. "Classic Coroutines" on page 645 introduced the concept, and Chapter 21 covers the use of native coroutines. Python coroutines usually run within a single thread under the supervision of an *event loop*, also in the same thread. Asynchronous programming frameworks such as *asyncio*, *Curio*, or *Trio* provide an event loop and supporting libraries for nonblocking, coroutine-based I/O. Coroutines support *cooperative multitasking*: each coroutine must explicitly cede control with the `yield` or `await` keyword, so that another may proceed concurrently (but not in parallel). This means that any blocking code in a coroutine blocks the execution of the event loop and all other coroutines—in contrast with the *preemptive multitasking* supported by processes and threads. On the other hand, each coroutine consumes less resources than a thread or process doing the same job.

Queue

A data structure that lets us put and get items, usually in FIFO order: first in, first out. Queues allow separate execution units to exchange application data and control messages, such as error codes and signals to terminate. The implementation of a queue varies according to the underlying concurrency model: the `queue` package in Python's standard library provides queue classes to support threads, while the multiprocessing and asyncio packages implement their own queue classes. The queue and asyncio packages also include queues that are not FIFO: LifoQueue and PriorityQueue.

Lock

An object that execution units can use to synchronize their actions and avoid corrupting data. While updating a shared data structure, the running code should hold an associated lock. This signals other parts of the program to wait until the lock is released before accessing the same data structure. The simplest type of lock is also known as a mutex (for mutual exclusion). The implementation of a lock depends on the underlying concurrency model.

Contention

Dispute over a limited asset. Resource contention happens when multiple execution units try to access a shared resource—such as a lock or storage. There's also CPU contention, when compute-intensive processes or threads must wait for the OS scheduler to give them a share of the CPU time.

Now let's use some of that jargon to understand concurrency support in Python.

Processes, Threads, and Python's Infamous GIL

Here is how the concepts we just saw apply to Python programming, in 10 points:

1. Each instance of the Python interpreter is a process. You can start additional Python processes using the *multiprocessing* or *concurrent.futures* libraries. Python's *subprocess* library is designed to launch processes to run external programs, regardless of the languages used to write them.

2. The Python interpreter uses a single thread to run the user's program and the memory garbage collector. You can start additional Python threads using the *threading* or *concurrent.futures* libraries.

3. Access to object reference counts and other internal interpreter state is controlled by a lock, the Global Interpreter Lock (GIL). Only one Python thread can hold the GIL at any time. This means that only one thread can execute Python code at any time, regardless of the number of CPU cores.

4. To prevent a Python thread from holding the GIL indefinitely, Python's bytecode interpreter pauses the current Python thread every 5ms by default,[4] releasing the

---

4 Call sys.getswitchinterval() (*https://fpy.li/19-3*) to get the interval; change it with sys.setswitchin terval(s) (*https://fpy.li/19-4*).

GIL. The thread can then try to reacquire the GIL, but if there are other threads waiting for it, the OS scheduler may pick one of them to proceed.

5. When we write Python code, we have no control over the GIL. But a built-in function or an extension written in C—or any language that interfaces at the Python/C API level—can release the GIL while running time-consuming tasks.

6. Every Python standard library function that makes a syscall[5] releases the GIL. This includes all functions that perform disk I/O, network I/O, and time.sleep(). Many CPU-intensive functions in the NumPy/SciPy libraries, as well as the compressing/decompressing functions from the zlib and bz2 modules, also release the GIL.[6]

7. Extensions that integrate at the Python/C API level can also launch other non-Python threads that are not affected by the GIL. Such GIL-free threads generally cannot change Python objects, but they can read from and write to the memory underlying objects that support the buffer protocol (*https://fpy.li/pep3118*), such as bytearray, array.array, and *NumPy* arrays.

8. The effect of the GIL on network programming with Python threads is relatively small, because the I/O functions release the GIL, and reading or writing to the network always implies high latency—compared to reading and writing to memory. Consequently, each individual thread spends a lot of time waiting anyway, so their execution can be interleaved without major impact on the overall throughput. That's why David Beazley says: "Python threads are great at doing nothing."[7]

9. Contention over the GIL slows down compute-intensive Python threads. Sequential, single-threaded code is simpler and faster for such tasks.

10. To run CPU-intensive Python code on multiple cores, you must use multiple Python processes.

Here is a good summary from the threading module documentation:[8]

> **CPython implementation detail**: In CPython, due to the Global Interpreter Lock, only one thread can execute Python code at once (even though certain performance-oriented libraries might overcome this limitation). If you want your application to

---

5 A syscall is a call from user code to a function of the operating system kernel. I/O, timers, and locks are some of the kernel services available through syscalls. To learn more, read the Wikipedia "System call" article (*https://fpy.li/19-5*).

6 The zlib and bz2 modules are specifically mentioned in a python-dev message by Antoine Pitrou (*https://fpy.li/19-6*), who contributed the time-slicing GIL logic to Python 3.2.

7 Source: slide 106 of Beazley's "Generators: The Final Frontier" tutorial (*https://fpy.li/19-7*).

8 Source: last paragraph of the "Thread objects" section (*https://fpy.li/19-8*).

make better use of the computational resources of multicore machines, you are advised to use multiprocessing or concurrent.futures.ProcessPoolExecutor. However, threading is still an appropriate model if you want to run multiple I/O-bound tasks simultaneously.

The previous paragraph starts with "CPython implementation detail" because the GIL is not part of the Python language definition. The Jython and IronPython implementations don't have a GIL. Unfortunately, both are lagging behind—still tracking Python 2.7. The highly performant PyPy interpreter (*https://fpy.li/19-9*) also has a GIL in its 2.7 and 3.7 versions—the latest as of June 2021.

This section did not mention coroutines, because by default they share the same Python thread among themselves and with the supervising event loop provided by an asynchronous framework, therefore the GIL does not affect them. It is possible to use multiple threads in an asynchronous program, but the best practice is that one thread runs the event loop and all coroutines, while additional threads carry out specific tasks. This will be explained in "Delegating Tasks to Executors" on page 801.

Enough concepts for now. Let's see some code.

A Concurrent Hello World

During a discussion about threads and how to avoid the GIL, Python contributor Michele Simionato posted an example (*https://fpy.li/19-10*) that is like a concurrent "Hello World": the simplest program to show how Python can "walk and chew gum."

Simionato's program uses multiprocessing, but I adapted it to introduce threading and asyncio as well. Let's start with the threading version, which may look familiar if you've studied threads in Java or C.

Spinner with Threads

The idea of the next few examples is simple: start a function that blocks for 3 seconds while animating characters in the terminal to let the user know that the program is "thinking" and not stalled.

The script makes an animated spinner displaying each character in the string "\|/-" in the same screen position.[9] When the slow computation finishes, the line with the spinner is cleared and the result is shown: Answer: 42.

---

9 Unicode has lots of characters useful for simple animations, like the Braille patterns (*https://fpy.li/19-11*) for example. I used the ASCII characters "\|/-" to keep the examples simple.

Figure 19-1 shows the output of two versions of the spinning example: first with threads, then with coroutines. If you're away from the computer, imagine the \ in the last line is spinning.

```
● ● ●                        ▦ 19-concurrency — Python spinner_async.py — 88×9
$ python3 spinner_thread.py
spinner object: <Thread(Thread-1 (spin), initial)>
Answer: 42
$ python3 spinner_async.py
spinner object: <Task pending name='Task-2' coro=<spin() running at /Users/luciano/flupy
/example-code-2e/19-concurrency/spinner_async.py:11>>
- thinking!█
```

Figure 19-1. The scripts spinner_thread.py and spinner_async.py produce similar output: the repr of a spinner object and the text "Answer: 42". In the screenshot, spinner_async.py is still running, and the animated message "/ thinking!" is shown; that line will be replaced by "Answer: 42" after 3 seconds.

Let's review the *spinner_thread.py* script first. Example 19-1 lists the first two functions in the script, and Example 19-2 shows the rest.

Example 19-1. spinner_thread.py: the spin and slow functions

```
import itertools
import time
from threading import Thread, Event

def spin(msg: str, done: Event) -> None:  ❶
    for char in itertools.cycle(r'\|/-'):  ❷
        status = f'\r{char} {msg}'  ❸
        print(status, end='', flush=True)
        if done.wait(.1):  ❹
            break  ❺
    blanks = ' ' * len(status)
    print(f'\r{blanks}\r', end='')  ❻

def slow() -> int:
    time.sleep(3)  ❼
    return 42
```

❶ This function will run in a separate thread. The done argument is an instance of threading.Event, a simple object to synchronize threads.

❷ This is an infinite loop because itertools.cycle yields one character at a time, cycling through the string forever.

❸ The trick for text-mode animation: move the cursor back to the start of the line with the carriage return ASCII control character ('\r').

❹ The `Event.wait(timeout=None)` method returns `True` when the event is set by another thread; if the `timeout` elapses, it returns `False`. The .1s timeout sets the "frame rate" of the animation to 10 FPS. If you want the spinner to go faster, use a smaller timeout.

❺ Exit the infinite loop.

❻ Clear the status line by overwriting with spaces and moving the cursor back to the beginning.

❼ `slow()` will be called by the main thread. Imagine this is a slow API call over the network. Calling `sleep` blocks the main thread, but the GIL is released so the spinner thread can proceed.

 The first important insight of this example is that `time.sleep()` blocks the calling thread but releases the GIL, allowing other Python threads to run.

The `spin` and `slow` functions will execute concurrently. The main thread—the only thread when the program starts—will start a new thread to run `spin` and then call `slow`. By design, there is no API for terminating a thread in Python. You must send it a message to shut down.

The `threading.Event` class is Python's simplest signalling mechanism to coordinate threads. An `Event` instance has an internal boolean flag that starts as `False`. Calling `Event.set()` sets the flag to `True`. While the flag is false, if a thread calls `Event.wait()`, it is blocked until another thread calls `Event.set()`, at which time `Event.wait()` returns `True`. If a timeout in seconds is given to `Event.wait(s)`, this call returns `False` when the timeout elapses, or returns `True` as soon as `Event.set()` is called by another thread.

The `supervisor` function, listed in Example 19-2, uses an `Event` to signal the `spin` function to exit.

Example 19-2. spinner_thread.py: the supervisor and main functions

```
def supervisor() -> int:  ❶
    done = Event()  ❷
    spinner = Thread(target=spin, args=('thinking!', done))  ❸
    print(f'spinner object: {spinner}')  ❹
    spinner.start()  ❺
    result = slow()  ❻
    done.set()  ❼
```

```
        spinner.join()  ❽
        return result

def main() -> None:
    result = supervisor()  ❾
    print(f'Answer: {result}')

if __name__ == '__main__':
    main()
```

❶ supervisor will return the result of slow.

❷ The threading.Event instance is the key to coordinate the activities of the main thread and the spinner thread, as explained further down.

❸ To create a new Thread, provide a function as the target keyword argument, and positional arguments to the target as a tuple passed via args.

❹ Display the spinner object. The output is <Thread(Thread-1, initial)>, where initial is the state of the thread—meaning it has not started.

❺ Start the spinner thread.

❻ Call slow, which blocks the main thread. Meanwhile, the secondary thread is running the spinner animation.

❼ Set the Event flag to True; this will terminate the for loop inside the spin function.

❽ Wait until the spinner thread finishes.

❾ Run the supervisor function. I wrote separate main and supervisor functions to make this example look more like the asyncio version in Example 19-4.

When the main thread sets the done event, the spinner thread will eventually notice and exit cleanly.

Now let's take a look at a similar example using the multiprocessing package.

Spinner with Processes

The multiprocessing package supports running concurrent tasks in separate Python processes instead of threads. When you create a multiprocessing.Process instance, a whole new Python interpreter is started as a child process in the background. Since each Python process has its own GIL, this allows your program to use all available

CPU cores—but that ultimately depends on the operating system scheduler. We'll see practical effects in "A Homegrown Process Pool" on page 720, but for this simple program it makes no real difference.

The point of this section is to introduce multiprocessing and show that its API emulates the threading API, making it easy to convert simple programs from threads to processes, as shown in *spinner_proc.py* (Example 19-3).

Example 19-3. spinner_proc.py: only the changed parts are shown; everything else is the same as spinner_thread.py

```
import itertools
import time
from multiprocessing import Process, Event     ❶
from multiprocessing import synchronize         ❷

def spin(msg: str, done: synchronize.Event) -> None:    ❸

# [snip] the rest of spin and slow functions are unchanged from spinner_thread.py

def supervisor() -> int:
    done = Event()
    spinner = Process(target=spin,                  ❹
                      args=('thinking!', done))
    print(f'spinner object: {spinner}')            ❺
    spinner.start()
    result = slow()
    done.set()
    spinner.join()
    return result

# [snip] main function is unchanged as well
```

❶ The basic multiprocessing API imitates the threading API, but type hints and Mypy expose this difference: multiprocessing.Event is a function (not a class like threading.Event) which returns a synchronize.Event instance...

❷ ...forcing us to import multiprocessing.synchronize...

❸ ...to write this type hint.

❹ Basic usage of the Process class is similar to Thread.

❺ The spinner object is displayed as <Process name='Process-1' parent=14868 initial>, where 14868 is the process ID of the Python instance running *spinner_proc.py*.

The basic API of threading and multiprocessing are similar, but their implementation is very different, and multiprocessing has a much larger API to handle the added complexity of multiprocess programming. For example, one challenge when converting from threads to processes is how to communicate between processes that are isolated by the operating system and can't share Python objects. This means that objects crossing process boundaries have to be serialized and deserialized, which creates overhead. In Example 19-3, the only data that crosses the process boundary is the Event state, which is implemented with a low-level OS semaphore in the C code underlying the multiprocessing module.[10]

Since Python 3.8, there's a multiprocessing.shared_memory (*https://fpy.li/19-12*) package in the standard library, but it does not support instances of user-defined classes. Besides raw bytes, the package allows processes to share a ShareableList, a mutable sequence type that can hold a fixed number of items of types int, float, bool, and None, as well as str and bytes up to 10 MB per item. See the ShareableList (*https://fpy.li/19-13*) documentation for more.

Now let's see how the same behavior can be achieved with coroutines instead of threads or processes.

Spinner with Coroutines

Chapter 21 is entirely devoted to asynchronous programming with coroutines. This is just a high-level introduction to contrast this approach with the threads and processes concurrency models. As such, we will overlook many details.

It is the job of OS schedulers to allocate CPU time to drive threads and processes. In contrast, coroutines are driven by an application-level event loop that manages a queue of pending coroutines, drives them one by one, monitors events triggered by I/O operations initiated by coroutines, and passes control back to the corresponding coroutine when each event happens. The event loop and the library coroutines and the user coroutines all execute in a single thread. Therefore, any time spent in a coroutine slows down the event loop—and all other coroutines.

---

10 The semaphore is a fundamental building block that can be used to implement other synchronization mechanisms. Python provides different semaphore classes for use with threads, processes, and coroutines. We'll see asyncio.Semaphore in "Using asyncio.as_completed and a Thread" on page 793 (Chapter 21).

The coroutine version of the spinner program is easier to understand if we start from the main function, then study the supervisor. That's what Example 19-4 shows.

Example 19-4. spinner_async.py: the main function and supervisor coroutine

```
def main() -> None:  ❶
    result = asyncio.run(supervisor())  ❷
    print(f'Answer: {result}')

async def supervisor() -> int:  ❸
    spinner = asyncio.create_task(spin('thinking!'))  ❹
    print(f'spinner object: {spinner}')  ❺
    result = await slow()  ❻
    spinner.cancel()  ❼
    return result

if __name__ == '__main__':
    main()
```

❶ main is the only regular function defined in this program—the others are coroutines.

❷ The asyncio.run function starts the event loop to drive the coroutine that will eventually set the other coroutines in motion. The main function will stay blocked until supervisor returns. The return value of supervisor will be the return value of asyncio.run.

❸ Native coroutines are defined with async def.

❹ asyncio.create_task schedules the eventual execution of spin, immediately returning an instance of asyncio.Task.

❺ The repr of the spinner object looks like <Task pending name='Task-2' coro=<spin() running at /path/to/spinner_async.py:11>>.

❻ The await keyword calls slow, blocking supervisor until slow returns. The return value of slow will be assigned to result.

❼ The Task.cancel method raises a CancelledError exception inside the spin coroutine, as we'll see in Example 19-5.

Example 19-4 demonstrates the three main ways of running a coroutine:

`asyncio.run(coro())`

Called from a regular function to drive a coroutine object that usually is the entry point for all the asynchronous code in the program, like the `supervisor` in this example. This call blocks until the body of `coro` returns. The return value of the `run()` call is whatever the body of `coro` returns.

`asyncio.create_task(coro())`

Called from a coroutine to schedule another coroutine to execute eventually. This call does not suspend the current coroutine. It returns a `Task` instance, an object that wraps the coroutine object and provides methods to control and query its state.

`await coro()`

Called from a coroutine to transfer control to the coroutine object returned by `coro()`. This suspends the current coroutine until the body of `coro` returns. The value of the await expression is whatever the body of `coro` returns.

 Remember: invoking a coroutine as `coro()` immediately returns a coroutine object, but does not run the body of the `coro` function. Driving the body of coroutines is the job of the event loop.

Now let's study the `spin` and `slow` coroutines in Example 19-5.

Example 19-5. spinner_async.py: the spin and slow coroutines

```python
import asyncio
import itertools

async def spin(msg: str) -> None:        ❶
    for char in itertools.cycle(r'\|/-'):
        status = f'\r{char} {msg}'
        print(status, flush=True, end='')
        try:
            await asyncio.sleep(.1)       ❷
        except asyncio.CancelledError:    ❸
            break
    blanks = ' ' * len(status)
    print(f'\r{blanks}\r', end='')

async def slow() -> int:
    await asyncio.sleep(3)                ❹
    return 42
```

❶ We don't need the Event argument that was used to signal that slow had completed its job in *spinner_thread.py* (Example 19-1).

❷ Use await asyncio.sleep(.1) instead of time.sleep(.1), to pause without blocking other coroutines. See the experiment after this example.

❸ asyncio.CancelledError is raised when the cancel method is called on the Task controlling this coroutine. Time to exit the loop.

❹ The slow coroutine also uses await asyncio.sleep instead of time.sleep.

Experiment: Break the spinner for an insight

Here is an experiment I recommend to understand how *spinner_async.py* works. Import the time module, then go to the slow coroutine and replace the line await asyncio.sleep(3) with a call to time.sleep(3), like in Example 19-6.

Example 19-6. spinner_async.py: replacing await asyncio.sleep(3) with time.sleep(3)

```
async def slow() -> int:
    time.sleep(3)
    return 42
```

Watching the behavior is more memorable than reading about it. Go ahead, I'll wait.

When you run the experiment, this is what you see:

1. The spinner object is shown, similar to this: <Task pending name='Task-2' coro=<spin() running at /path/to/spinner_async.py:12>>.
2. The spinner never appears. The program hangs for 3 seconds.
3. Answer: 42 is displayed and the program ends.

To understand what is happening, recall that Python code using asyncio has only one flow of execution, unless you've explicitly started additional threads or processes. That means only one coroutine executes at any point in time. Concurrency is achieved by control passing from one coroutine to another. In Example 19-7, let's focus on what happens in the supervisor and slow coroutines during the proposed experiment.

Example 19-7. spinner_async_experiment.py: the supervisor and slow coroutines

```
async def slow() -> int:
    time.sleep(3)  ❹
```

```
    return 42

async def supervisor() -> int:
    spinner = asyncio.create_task(spin('thinking!'))  ❶
    print(f'spinner object: {spinner}')  ❷
    result = await slow()  ❸
    spinner.cancel()  ❺
    return result
```

❶ The `spinner` task is created, to eventually drive the execution of `spin`.

❷ The display shows the `Task` is "pending."

❸ The `await` expression transfers control to the `slow` coroutine.

❹ `time.sleep(3)` blocks for 3 seconds; nothing else can happen in the program, because the main thread is blocked—and it is the only thread. The operating system will continue with other activities. After 3 seconds, `sleep` unblocks, and `slow` returns.

❺ Right after `slow` returns, the `spinner` task is cancelled. The flow of control never reached the body of the `spin` coroutine.

The *spinner_async_experiment.py* teaches an important lesson, as explained in the following warning.

 Never use `time.sleep(…)` in `asyncio` coroutines unless you want to pause your whole program. If a coroutine needs to spend some time doing nothing, it should `await asyncio.sleep(DELAY)`. This yields control back to the `asyncio` event loop, which can drive other pending coroutines.

Greenlet and gevent

As we discuss concurrency with coroutines, it's important to mention the *greenlet* (*https://fpy.li/19-14*) package, which has been around for many years and is used at scale.[11] The package supports cooperative multitasking through lightweight coroutines—named *greenlets*—that don't require any special syntax such as `yield` or `await`, therefore are easier to integrate into existing, sequential codebases. SQL Alchemy 1.4 ORM uses greenlets (*https://fpy.li/19-15*) internally to implement its new asynchronous API (*https://fpy.li/19-16*) compatible with *asyncio*.

---

11 Thanks to tech reviewers Caleb Hattingh and Jürgen Gmach who did not let me overlook *greenlet* and *gevent*.

The *gevent* (*https://fpy.li/19-17*) networking library monkey patches Python's standard socket module making it nonblocking by replacing some of its code with greenlets. To a large extent, *gevent* is transparent to the surrounding code, making it easier to adapt sequential applications and libraries—such as database drivers—to perform concurrent network I/O. Numerous open source projects (*https://fpy.li/19-18*) use *gevent*, including the widely deployed *Gunicorn* (*https://fpy.li/gunicorn*)—mentioned in "WSGI Application Servers" on page 734.

Supervisors Side-by-Side

The line count of *spinner_thread.py* and *spinner_async.py* is nearly the same. The supervisor functions are the heart of these examples. Let's compare them in detail. Example 19-8 lists only the supervisor from Example 19-2.

Example 19-8. spinner_thread.py: the threaded supervisor function

```
def supervisor() -> int:
    done = Event()
    spinner = Thread(target=spin,
                     args=('thinking!', done))
    print('spinner object:', spinner)
    spinner.start()
    result = slow()
    done.set()
    spinner.join()
    return result
```

For comparison, Example 19-9 shows the supervisor coroutine from Example 19-4.

Example 19-9. spinner_async.py: the asynchronous supervisor coroutine

```
async def supervisor() -> int:
    spinner = asyncio.create_task(spin('thinking!'))
    print('spinner object:', spinner)
    result = await slow()
    spinner.cancel()
    return result
```

Here is a summary of the differences and similarities to note between the two super visor implementations:

- An asyncio.Task is roughly the equivalent of a threading.Thread.
- A Task drives a coroutine object, and a Thread invokes a callable.
- A coroutine yields control explicitly with the await keyword.

- You don't instantiate `Task` objects yourself, you get them by passing a coroutine to `asyncio.create_task(…)`.

- When `asyncio.create_task(…)` returns a `Task` object, it is already scheduled to run, but a `Thread` instance must be explicitly told to run by calling its `start` method.

- In the threaded `supervisor`, `slow` is a plain function and is directly invoked by the main thread. In the asynchronous `supervisor`, `slow` is a coroutine driven by `await`.

- There's no API to terminate a thread from the outside; instead, you must send a signal—like setting the `done` Event object. For tasks, there is the `Task.cancel()` instance method, which raises `CancelledError` at the `await` expression where the coroutine body is currently suspended.

- The `supervisor` coroutine must be started with `asyncio.run` in the `main` function.

This comparison should help you understand how concurrent jobs are orchestrated with *asyncio*, in contrast to how it's done with the `Threading` module, which may be more familiar to you.

One final point related to threads versus coroutines: if you've done any nontrivial programming with threads, you know how challenging it is to reason about the program because the scheduler can interrupt a thread at any time. You must remember to hold locks to protect the critical sections of your program, to avoid getting interrupted in the middle of a multistep operation—which could leave data in an invalid state.

With coroutines, your code is protected against interruption by default. You must explicitly `await` to let the rest of the program run. Instead of holding locks to synchronize the operations of multiple threads, coroutines are "synchronized" by definition: only one of them is running at any time. When you want to give up control, you use `await` to yield control back to the scheduler. That's why it is possible to safely cancel a coroutine: by definition, a coroutine can only be cancelled when it's suspended at an `await` expression, so you can perform cleanup by handling the `Cancel ledError` exception.

The `time.sleep()` call blocks but does nothing. Now we'll experiment with a CPU-intensive call to get a better understanding of the GIL, as well as the effect of CPU-intensive functions in asynchronous code.

The Real Impact of the GIL

In the threading code (Example 19-1), you can replace the time.sleep(3) call in the slow function with an HTTP client request from your favorite library, and the spinner will keep spinning. That's because a well-designed network library will release the GIL while waiting for the network.

You can also replace the asyncio.sleep(3) expression in the slow coroutine to await for a response from a well-designed asynchronous network library, because such libraries provide coroutines that yield control back to the event loop while waiting for the network. Meanwhile, the spinner will keep spinning.

With CPU-intensive code, the story is different. Consider the function is_prime in Example 19-10, which returns True if the argument is a prime number, False if it's not.

Example 19-10. primes.py: an easy to read primality check, from Python's ProcessPool Executor example (https://fpy.li/19-19)

```
def is_prime(n: int) -> bool:
    if n < 2:
        return False
    if n == 2:
        return True
    if n % 2 == 0:
        return False

    root = math.isqrt(n)
    for i in range(3, root + 1, 2):
        if n % i == 0:
            return False
    return True
```

The call is_prime(5_000_111_000_222_021) takes about 3.3s on the company laptop I am using now.[12]

Quick Quiz

Given what we've seen so far, please take the time to consider the following three-part question. One part of the answer is tricky (at least it was for me).

> What would happen to the spinner animation if you made the following changes, assuming that n = 5_000_111_000_222_021—that prime which my machine takes 3.3s to verify:

---

12 It's a 15" MacBook Pro 2018 with a 6-core, 2.2 GHz Intel Core i7 CPU.

1. In *spinner_proc.py*, replace `time.sleep(3)` with a call to `is_prime(n)`?

2. In *spinner_thread.py*, replace `time.sleep(3)` with a call to `is_prime(n)`?

3. In *spinner_async.py*, replace `await asyncio.sleep(3)` with a call to `is_prime(n)`?

Before you run the code or read on, I recommend figuring out the answers on your own. Then, you may want to copy and modify the *spinner_*.py* examples as suggested.

Now the answers, from easier to hardest.

1. Answer for multiprocessing

The spinner is controlled by a child process, so it continues spinning while the primality test is computed by the parent process.[13]

2. Answer for threading

The spinner is controlled by a secondary thread, so it continues spinning while the primality test is computed by the main thread.

I did not get this answer right at first: I was expecting the spinner to freeze because I overestimated the impact of the GIL.

In this particular example, the spinner keeps spinning because Python suspends the running thread every 5ms (by default), making the GIL available to other pending threads. Therefore, the main thread running `is_prime` is interrupted every 5ms, allowing the secondary thread to wake up and iterate once through the `for` loop, until it calls the `wait` method of the `done` event, at which time it will release the GIL. The main thread will then grab the GIL, and the `is_prime` computation will proceed for another 5ms.

This does not have a visible impact on the running time of this specific example, because the `spin` function quickly iterates once and releases the GIL as it waits for the `done` event, so there is not much contention for the GIL. The main thread running `is_prime` will have the GIL most of the time.

We got away with a compute-intensive task using threading in this simple experiment because there are only two threads: one hogging the CPU, and the other waking up only 10 times per second to update the spinner.

---

13 This is true today because you are probably using a modern OS with *preemptive multitasking*. Windows before the NT era and macOS before the OSX era were not "preemptive," therefore any process could take over 100% of the CPU and freeze the whole system. We are not completely free of this kind of problem today but trust this graybeard: this troubled every user in the 1990s, and a hard reset was the only cure.

But if you have two or more threads vying for a lot of CPU time, your program will be slower than sequential code.

3. Answer for asyncio

If you call is_prime(5_000_111_000_222_021) in the slow coroutine of the *spinner_async.py* example, the spinner will never appear. The effect would be the same we had in Example 19-6, when we replaced await asyncio.sleep(3) with time.sleep(3): no spinning at all. The flow of control will pass from supervisor to slow, and then to is_prime. When is_prime returns, slow returns as well, and supervisor resumes, cancelling the spinner task before it is executed even once. The program appears frozen for about 3s, then shows the answer.

Power Napping with sleep(0)

One way to keep the spinner alive is to rewrite is_prime as a coroutine, and periodically call asyncio.sleep(0) in an await expression to yield control back to the event loop, like in Example 19-11.

Example 19-11. spinner_async_nap.py: is_prime is now a coroutine

```
async def is_prime(n):
    if n < 2:
        return False
    if n == 2:
        return True
    if n % 2 == 0:
        return False

    root = math.isqrt(n)
    for i in range(3, root + 1, 2):
        if n % i == 0:
            return False
        if i % 100_000 == 1:
            await asyncio.sleep(0)    ❶
    return True
```

❶ Sleep once every 50,000 iterations (because the step in the range is 2).

Issue #284 (*https://fpy.li/19-20*) in the asyncio repository has an informative discussion about the use of asyncio.sleep(0).

However, be aware this will slow down is_prime, and—more importantly—will still slow down the event loop and your whole program with it. When I used await asyncio.sleep(0) every 100,000 iterations, the spinner was smooth but the program ran in 4.9s on my machine, almost 50% longer than the original primes.is_prime function by itself with the same argument (5_000_111_000_222_021).

Using `await asyncio.sleep(0)` should be considered a stopgap measure before you refactor your asynchronous code to delegate CPU-intensive computations to another process. We'll see one way of doing that with `asyncio.loop.run_in_executor` (*https://fpy.li/19-21*), covered in Chapter 21. Another option would be a task queue, which we'll briefly discuss in "Distributed Task Queues" on page 736.

So far, we've only experimented with a single call to a CPU-intensive function. The next section presents concurrent execution of multiple CPU-intensive calls.

A Homegrown Process Pool

 I wrote this section to show the use of multiple processes for CPU-intensive tasks, and the common pattern of using queues to distribute tasks and collect results. Chapter 20 will show a simpler way of distributing tasks to processes: a `ProcessPoolExecutor` from the `concurrent.futures` package, which uses queues internally.

In this section we'll write programs to compute the primality of a sample of 20 integers, from 2 to 9,999,999,999,999,999—i.e., $10^{16} - 1$, or more than 2^{53}. The sample includes small and large primes, as well as composite numbers with small and large prime factors.

The *sequential.py* program provides the performance baseline. Here is a sample run:

```
$ python3 sequential.py
               2  P  0.000001s
 142702110479723  P  0.568328s
 299593572317531  P  0.796773s
3333333333333301  P  2.648625s
3333333333333333     0.000007s
3333335652092209     2.672323s
4444444444444423  P  3.052667s
4444444444444444     0.000001s
4444444488888889     3.061083s
5555553133149889     3.451833s
5555555555555503  P  3.556867s
5555555555555555     0.000007s
6666666666666666     0.000001s
6666666666666719  P  3.781064s
6666667141414921     3.778166s
7777777536340681     4.120069s
7777777777777753  P  4.141530s
7777777777777777     0.000007s
9999999999999917  P  4.678164s
9999999999999999     0.000007s
Total time: 40.31
```

The results are shown in three columns:

- The number to be checked.
- P if it's a prime number, blank if not.
- Elapsed time for checking the primality for that specific number.

In this example, the total time is approximately the sum of the times for each check, but it is computed separately, as you can see in Example 19-12.

Example 19-12. sequential.py: sequential primality check for a small dataset

```python
#!/usr/bin/env python3

"""
sequential.py: baseline for comparing sequential, multiprocessing,
and threading code for CPU-intensive work.
"""

from time import perf_counter
from typing import NamedTuple

from primes import is_prime, NUMBERS

class Result(NamedTuple):    ❶
    prime: bool
    elapsed: float

def check(n: int) -> Result:    ❷
    t0 = perf_counter()
    prime = is_prime(n)
    return Result(prime, perf_counter() - t0)

def main() -> None:
    print(f'Checking {len(NUMBERS)} numbers sequentially:')
    t0 = perf_counter()
    for n in NUMBERS:    ❸
        prime, elapsed = check(n)
        label = 'P' if prime else ' '
        print(f'{n:16}  {label} {elapsed:9.6f}s')

    elapsed = perf_counter() - t0    ❹
    print(f'Total time: {elapsed:.2f}s')

if __name__ == '__main__':
    main()
```

❶ The check function (in the next callout) returns a Result tuple with the boolean value of the is_prime call and the elapsed time.

❷ check(n) calls is_prime(n) and computes the elapsed time to return a Result.

❸ For each number in the sample, we call check and display the result.

❹ Compute and display the total elapsed time.

Process-Based Solution

The next example, *procs.py*, shows the use of multiple processes to distribute the primality checks across multiple CPU cores. These are the times I get with *procs.py*:

```
$ python3 procs.py
Checking 20 numbers with 12 processes:
              2  P  0.000002s
3333333333333333     0.000021s
4444444444444444     0.000002s
5555555555555555     0.000018s
6666666666666666     0.000002s
 142702110479723  P  1.350982s
7777777777777777     0.000009s
 299593572317531  P  1.981411s
9999999999999999     0.000008s
3333333333333301  P  6.328173s
3333335652092209     6.419249s
4444444488888889     7.051267s
4444444444444423  P  7.122004s
5555553133149889     7.412735s
5555555555555503  P  7.603327s
6666666666666719  P  7.934670s
6666667141414921     8.017599s
7777777536340681     8.339623s
7777777777777753  P  8.388859s
9999999999999917  P  8.117313s
20 checks in 9.58s
```

The last line of the output shows that *procs.py* was 4.2 times faster than *sequential.py*.

Understanding the Elapsed Times

Note that the elapsed time in the first column is for checking that specific number. For example, is_prime(7777777777777753) took almost 8.4s to return True. Meanwhile, other processes were checking other numbers in parallel.

There were 20 numbers to check. I wrote *procs.py* to start a number of worker processes equal to the number of CPU cores, as determined by multiprocessing.cpu_count().

The total time in this case is much less than the sum of the elapsed time for the individual checks. There is some overhead in spinning up processes and in inter-process communication, so the end result is that the multiprocess version is only about 4.2 times faster than the sequential. That's good, but a little disappointing considering the code launches 12 processes to use all cores on this laptop.

 The `multiprocessing.cpu_count()` function returns 12 on the MacBook Pro I'm using to write this chapter. It's actually a 6-CPU Core-i7, but the OS reports 12 CPUs because of hyperthreading, an Intel technology which executes 2 threads per core. However, hyperthreading works better when one of the threads is not working as hard as the other thread in the same core—perhaps the first is stalled waiting for data after a cache miss, and the other is crunching numbers. Anyway, there's no free lunch: this laptop performs like a 6-CPU machine for compute-intensive work that doesn't use a lot of memory—like that simple primality test.

Code for the Multicore Prime Checker

When we delegate computing to threads or processes, our code does not call the worker function directly, so we can't simply get a return value. Instead, the worker is driven by the thread or process library, and it eventually produces a result that needs to be stored somewhere. Coordinating workers and collecting results are common uses of queues in concurrent programming—and also in distributed systems.

Much of the new code in *procs.py* has to do with setting up and using queues. The top of the file is in Example 19-13.

 `SimpleQueue` was added to `multiprocessing` in Python 3.9. If you're using an earlier version of Python, you can replace `Simple Queue` with `Queue` in Example 19-13.

Example 19-13. procs.py: multiprocess primality check; imports, types, and functions

```
import sys
from time import perf_counter
from typing import NamedTuple
from multiprocessing import Process, SimpleQueue, cpu_count   ❶
from multiprocessing import queues   ❷

from primes import is_prime, NUMBERS

class PrimeResult(NamedTuple):   ❸
    n: int
```

```
        prime: bool
        elapsed: float

JobQueue = queues.SimpleQueue[int]   ❹
ResultQueue = queues.SimpleQueue[PrimeResult]   ❺

def check(n: int) -> PrimeResult:   ❻
    t0 = perf_counter()
    res = is_prime(n)
    return PrimeResult(n, res, perf_counter() - t0)

def worker(jobs: JobQueue, results: ResultQueue) -> None:   ❼
    while n := jobs.get():   ❽
        results.put(check(n))   ❾
    results.put(PrimeResult(0, False, 0.0))   ❿

def start_jobs(
    procs: int, jobs: JobQueue, results: ResultQueue   ⓫
) -> None:
    for n in NUMBERS:
        jobs.put(n)   ⓬
    for _ in range(procs):
        proc = Process(target=worker, args=(jobs, results))   ⓭
        proc.start()   ⓮
        jobs.put(0)   ⓯
```

❶ Trying to emulate threading, multiprocessing provides multiprocessing.Sim pleQueue, but this is a method bound to a predefined instance of a lower-level BaseContext class. We must call this SimpleQueue to build a queue, we can't use it in type hints.

❷ multiprocessing.queues has the SimpleQueue class we need for type hints.

❸ PrimeResult includes the number checked for primality. Keeping n together with the other result fields simplifies displaying results later.

❹ This is a type alias for a SimpleQueue that the main function (Example 19-14) will use to send numbers to the processes that will do the work.

❺ Type alias for a second SimpleQueue that will collect the results in main. The values in the queue will be tuples made of the number to be tested for primality, and a Result tuple.

❻ This is similar to *sequential.py*.

❼ worker gets a queue with the numbers to be checked, and another to put results.

❽ In this code, I use the number 0 as a *poison pill*: a signal for the worker to finish. If n is not 0, proceed with the loop.[14]

❾ Invoke the primality check and enqueue `PrimeResult`.

❿ Send back a `PrimeResult(0, False, 0.0)` to let the main loop know that this worker is done.

⓫ `procs` is the number of processes that will compute the prime checks in parallel.

⓬ Enqueue the numbers to be checked in `jobs`.

⓭ Fork a child process for each worker. Each child will run the loop inside its own instance of the `worker` function, until it fetches a 0 from the `jobs` queue.

⓮ Start each child process.

⓯ Enqueue one 0 for each process, to terminate them.

Loops, Sentinels, and Poison Pills

The `worker` function in Example 19-13 follows a common pattern in concurrent programming: looping indefinitely while taking items from a queue and processing each with a function that does the actual work. The loop ends when the queue produces a sentinel value. In this pattern, the sentinel that shuts down the worker is often called a "poison pill."

`None` is often used as a sentinel value, but it may be unsuitable if it can occur in the data stream. Calling `object()` is a common way to get a unique value to use as sentinel. However, that does not work across processes because Python objects must be serialized for inter-process communication, and when you `pickle.dump` and `pickle.load` an instance of `object`, the unpickled instance is distinct from the original: it doesn't compare equal. A good alternative to `None` is the `Ellipsis` built-in object (a.k.a. `...`), which survives serialization without losing its identity.[15]

Python's standard library uses lots of different values (*https://fpy.li/19-22*) as sentinels. PEP 661—Sentinel Values (*https://fpy.li/pep661*) proposes a standard sentinel type. As of September 2021, it's only a draft.

---

14 In this example, 0 is a convenient sentinel. None is also commonly used for that. Using 0 simplifies the type hint for PrimeResult and the code for worker.

15 Surviving serialization without losing our identity is a pretty good life goal.

Now let's study the `main` function of *procs.py* in Example 19-14.

Example 19-14. procs.py: multiprocess primality check; main function

```python
def main() -> None:
    if len(sys.argv) < 2:        ❶
        procs = cpu_count()
    else:
        procs = int(sys.argv[1])

    print(f'Checking {len(NUMBERS)} numbers with {procs} processes:')
    t0 = perf_counter()
    jobs: JobQueue = SimpleQueue()        ❷
    results: ResultQueue = SimpleQueue()
    start_jobs(procs, jobs, results)      ❸
    checked = report(procs, results)      ❹
    elapsed = perf_counter() - t0
    print(f'{checked} checks in {elapsed:.2f}s')        ❺

def report(procs: int, results: ResultQueue) -> int:    ❻
    checked = 0
    procs_done = 0
    while procs_done < procs:        ❼
        n, prime, elapsed = results.get()        ❽
        if n == 0:        ❾
            procs_done += 1
        else:
            checked += 1        ❿
            label = 'P' if prime else ' '
            print(f'{n:16}   {label} {elapsed:9.6f}s')
    return checked

if __name__ == '__main__':
    main()
```

❶ If no command-line argument is given, set the number of processes to the number of CPU cores; otherwise, create as many processes as given in the first argument.

❷ `jobs` and `results` are the queues described in Example 19-13.

❸ Start `proc` processes to consume `jobs` and post `results`.

❹ Retrieve the results and display them; `report` is defined in ❻.

❺ Display how many numbers were checked and the total elapsed time.

❻ The arguments are the number of `procs` and the queue to post the results.

❼ Loop until all processes are done.

❽ Get one `PrimeResult`. Calling `.get()` on a queue block until there is an item in the queue. It's also possible to make this nonblocking, or set a timeout. See the `SimpleQueue.get` (*https://fpy.li/19-23*) documentation for details.

❾ If n is zero, then one process exited; increment the `procs_done` count.

❿ Otherwise, increment the `checked` count (to keep track of the numbers checked) and display the results.

The results will not come back in the same order the jobs were submitted. That's why I had to put n in each `PrimeResult` tuple. Otherwise, I'd have no way to know which result belonged to each number.

If the main process exits before all subprocesses are done, you may see confusing tracebacks on `FileNotFoundError` exceptions caused by an internal lock in `multi processing`. Debugging concurrent code is always hard, and debugging `multiproc essing` is even harder because of all the complexity behind the thread-like façade. Fortunately, the `ProcessPoolExecutor` we'll meet in Chapter 20 is easier to use and more robust.

 Thanks to reader Michael Albert who noticed the code I published during the early release had a *race condition* (*https://fpy.li/19-24*) in Example 19-14. A race condition is a bug that may or may not occur depending on the order of actions performed by concurrent execution units. If "A" happens before "B," all is fine; but it "B" happens first, something goes wrong. That's the race.

If you are curious, this diff shows the bug and how I fixed it: *example-code-2e/commit/2c123057* (*https://fpy.li/19-25*)—but note that I later refactored the example to delegate parts of `main` to the `start_jobs` and `report` functions. There's a *README.md* (*https:// fpy.li/19-26*) file in the same directory explaining the problem and the solution.

Experimenting with More or Fewer Processes

You may want try running *procs.py*, passing arguments to set the number of worker processes. For example, this command…

```
$ python3 procs.py 2
```

…will launch two worker processes, producing results almost twice as fast as *sequential.py*—if your machine has at least two cores and is not too busy running other programs.

I ran *procs.py* 12 times with 1 to 20 processes, totaling 240 runs. Then I computed the median time for all runs with the same number of processes, and plotted Figure 19-2.

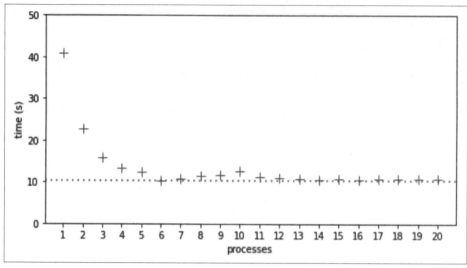

Figure 19-2. Median run times for each number of processes from 1 to 20. Highest median time was 40.81s, with 1 process. Lowest median time was 10.39s, with 6 processes, indicated by the dotted line.

In this 6-core laptop, the lowest median time was with 6 processes: 10.39s—marked by the dotted line in Figure 19-2. I expected the run time to increase after 6 processes due to CPU contention, and it reached a local maximum of 12.51s at 10 processes. I did not expect and I can't explain why the performance improved at 11 processes and remained almost flat from 13 to 20 processes, with median times only slightly higher than the lowest median time at 6 processes.

Thread-Based Nonsolution

I also wrote *threads.py*, a version of *procs.py* using `threading` instead of `multiproc essing`. The code is very similar—as is usually the case when converting simple examples between these two APIs.[16] Due to the GIL and the compute-intensive nature of `is_prime`, the threaded version is slower than the sequential code in Example 19-12, and it gets slower as the number of threads increase, because of CPU contention and the cost of context switching. To switch to a new thread, the OS needs to save CPU registers and update the program counter and stack pointer,

---

16 See *19-concurrency/primes/threads.py* (*https://fpy.li/19-27*) in the *Fluent Python* code repository (*https://fpy.li/ code*).

triggering expensive side effects like invalidating CPU caches and possibly even swapping memory pages.[17]

The next two chapters will cover more about concurrent programming in Python, using the high-level *concurrent.futures* library to manage threads and processes (Chapter 20) and the *asyncio* library for asynchronous programming (Chapter 21).

The remaining sections in this chapter aim to answer the question:

> Given the limitations discussed so far, how is Python thriving in a multicore world?

Python in the Multicore World

Consider this citation from the widely quoted article "The Free Lunch Is Over: A Fundamental Turn Toward Concurrency in Software" by Herb Sutter (*https://fpy.li/ 19-29*):

> The major processor manufacturers and architectures, from Intel and AMD to Sparc and PowerPC, have run out of room with most of their traditional approaches to boosting CPU performance. Instead of driving clock speeds and straight-line instruction throughput ever higher, they are instead turning en masse to hyper-threading and multicore architectures. March 2005. [Available online].

What Sutter calls the "free lunch" was the trend of software getting faster with no additional developer effort because CPUs were executing sequential code faster, year after year. Since 2004, that is no longer true: clock speeds and execution optimizations reached a plateau, and now any significant increase in performance must come from leveraging multiple cores or hyperthreading, advances that only benefit code that is written for concurrent execution.

Python's story started in the early 1990s, when CPUs were still getting exponentially faster at sequential code execution. There was no talk about multicore CPUs except in supercomputers back then. At the time, the decision to have a GIL was a no-brainer. The GIL makes the interpreter faster when running on a single core, and its implementation simpler.[18] The GIL also makes it easier to write simple extensions through the Python/C API.

---

17 To learn more, see "Context switch" (*https://fpy.li/19-28*) in the English Wikipedia.

18 These are probably the same reasons that prompted the creator of the Ruby language, Yukihiro Matsumoto, to use a GIL in his interpreter as well.

I just wrote "simple extensions" because an extension does not need to deal with the GIL at all. A function written in C or Fortran may be hundreds of times faster than the equivalent in Python.[19] Therefore the added complexity of releasing the GIL to leverage multicore CPUs may not be needed in many cases. So we can thank the GIL for many extensions available for Python—and that is certainly one of the key reasons why the language is so popular today.

Despite the GIL, Python is thriving in applications that require concurrent or parallel execution, thanks to libraries and software architectures that work around the limitations of CPython.

Now let's discuss how Python is used in system administration, data science, and server-side application development in the multicore, distributed computing world of 2021.

System Administration

Python is widely used to manage large fleets of servers, routers, load balancers, and network-attached storage (NAS). It's also a leading option in software-defined networking (SDN) and ethical hacking. Major cloud service providers support Python through libraries and tutorials authored by the providers themselves or by their large communities of Python users.

In this domain, Python scripts automate configuration tasks by issuing commands to be carried out by the remote machines, so rarely there are CPU-bound operations to be done. Threads or coroutines are well suited for such jobs. In particular, the concur rent.futures package we'll see in Chapter 20 can be used to perform the same operations on many remote machines at the same time without a lot of complexity.

Beyond the standard library, there are popular Python-based projects to manage server clusters: tools like *Ansible* (*https://fpy.li/19-30*) and *Salt* (*https://fpy.li/19-31*), as well as libraries like *Fabric* (*https://fpy.li/19-32*).

There is also a growing number of libraries for system administration supporting coroutines and asyncio. In 2016, Facebook's Production Engineering team reported (*https://fpy.li/19-33*): "We are increasingly relying on AsyncIO, which was introduced in Python 3.4, and seeing huge performance gains as we move codebases away from Python 2."

---

19 As an exercise in college, I had to implement the LZW compression algorithm in C. But first I wrote it in Python, to check my understanding of the spec. The C version was about 900× faster.

Data Science

Data science—including artificial intelligence—and scientific computing are very well served by Python. Applications in these fields are compute-intensive, but Python users benefit from a vast ecosystem of numeric computing libraries written in C, C++, Fortran, Cython, etc.—many of which are able to leverage multicore machines, GPUs, and/or distributed parallel computing in heterogeneous clusters.

As of 2021, Python's data science ecosystem includes impressive tools such as:

Project Jupyter (https://fpy.li/19-34)
> Two browser-based interfaces—Jupyter Notebook and JupyterLab—that allow users to run and document analytics code potentially running across the network on remote machines. Both are hybrid Python/JavaScript applications, supporting computing kernels written in different languages, all integrated via ZeroMQ—an asynchronous messaging library for distributed applications. The name *Jupyter* actually comes from Julia, Python, and R, the first three languages supported by the Notebook. The rich ecosystem built on top of the Jupyter tools include Bokeh (*https://fpy.li/19-35*), a powerful interactive visualization library that lets users navigate and interact with large datasets or continuously updated streaming data, thanks to the performance of modern JavaScript engines and browsers.

TensorFlow (https://fpy.li/19-36) and PyTorch (https://fpy.li/19-37)
> These are the top two deep learning frameworks, according to O'Reilly's January 2021 report (*https://fpy.li/19-38*) on usage of their learning resources during 2020. Both projects are written in C++, and are able to leverage multiple cores, GPUs, and clusters. They support other languages as well, but Python is their main focus and is used by the majority of their users. TensorFlow was created and is used internally by Google; PyTorch by Facebook.

Dask (https://fpy.li/dask)
> A parallel computing library that can farm out work to local processes or clusters of machines, "tested on some of the largest supercomputers in the world"—as their home page (*https://fpy.li/dask*) states. Dask offers APIs that closely emulate NumPy, pandas, and scikit-learn—the most popular libraries in data science and machine learning today. Dask can be used from JupyterLab or Jupyter Notebook, and leverages Bokeh not only for data visualization but also for an interactive dashboard showing the flow of data and computations across the processes/machines in near real time. Dask is so impressive that I recommend watching a video such as this 15-minute demo (*https://fpy.li/19-39*) in which Matthew Rocklin—a maintainer of the project—shows Dask crunching data on 64 cores distributed in 8 EC2 machines on AWS.

These are only some examples to illustrate how the data science community is creating solutions that leverage the best of Python and overcome the limitations of the CPython runtime.

Server-Side Web/Mobile Development

Python is widely used in web applications and for the backend APIs supporting mobile applications. How is it that Google, YouTube, Dropbox, Instagram, Quora, and Reddit—among others—managed to build Python server-side applications serving hundreds of millions of users 24x7? Again, the answer goes way beyond what Python provides "out of the box."

Before we discuss tools to support Python at scale, I must quote an admonition from the Thoughtworks *Technology Radar*:

> **High performance envy/web scale envy**
>
> We see many teams run into trouble because they have chosen complex tools, frameworks or architectures because they "might need to scale." Companies such as Twitter and Netflix need to support extreme loads and so need these architectures, but they also have extremely skilled development teams able to handle the complexity. Most situations do not require these kinds of engineering feats; teams should keep their *web scale envy* in check in favor of simpler solutions that still get the job done.[20]

At *web scale*, the key is an architecture that allows horizontal scaling. At that point, all systems are distributed systems, and no single programming language is likely to be the right choice for every part of the solution.

Distributed systems is a field of academic research, but fortunately some practitioners have written accessible books anchored on solid research and practical experience. One of them is Martin Kleppmann, the author of *Designing Data-Intensive Applications* (O'Reilly).

Consider Figure 19-3, the first of many architecture diagrams in Kleppmann's book. Here are some components I've seen in Python engagements that I worked on or have firsthand knowledge of:

- Application caches:[21] *memcached, Redis, Varnish*
- Relational databases: *PostgreSQL, MySQL*

---

20 Source: Thoughtworks Technology Advisory Board, *Technology Radar*—November 2015 (*https://fpy.li/19-40*).

21 Contrast application caches—used directly by your application code—with HTTP caches, which would be placed on the top edge of Figure 19-3 to serve static assets like images, CSS, and JS files. Content Delivery Networks (CDNs) offer another type of HTTP cache, deployed in data centers closer to the end users of your application.

- Document databases: *Apache CouchDB, MongoDB*
- Full-text indexes: *Elasticsearch, Apache Solr*
- Message queues: *RabbitMQ, Redis*

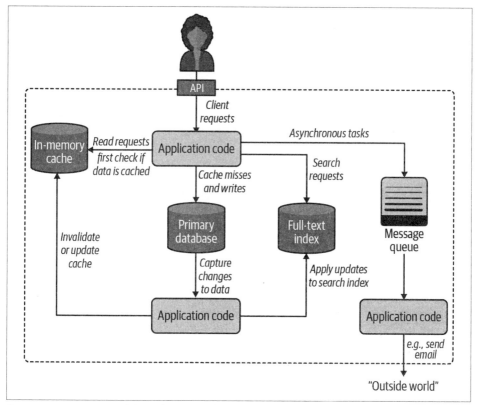

Figure 19-3. One possible architecture for a system combining several components.[22]

There are other industrial-strength open source products in each of those categories. Major cloud providers also offer their own proprietary alternatives.

Kleppmann's diagram is general and language independent—as is his book. For Python server-side applications, two specific components are often deployed:

- An application server to distribute the load among several instances of the Python application. The application server would appear near the top in Figure 19-3, handling client requests before they reached the application code.

---

22 Diagram adapted from Figure 1-1, *Designing Data-Intensive Applications* by Martin Kleppmann (O'Reilly).

- A task queue built around the message queue on the righthand side of Figure 19-3, providing a higher-level, easier-to-use API to distribute tasks to processes running on other machines.

The next two sections explore these components that are recommended best practices in Python server-side deployments.

WSGI Application Servers

WSGI—the Web Server Gateway Interface (*https://fpy.li/pep3333*)—is a standard API for a Python framework or application to receive requests from an HTTP server and send responses to it.[23] WSGI application servers manage one or more processes running your application, maximizing the use of the available CPUs.

Figure 19-4 illustrates a typical WSGI deployment.

 If we wanted to merge the previous pair of diagrams, the content of the dashed rectangle in Figure 19-4 would replace the solid "Application code" rectangle at the top of Figure 19-3.

The best-known application servers in Python web projects are:

- *mod_wsgi* (*https://fpy.li/19-41*)
- *uWSGI* (*https://fpy.li/19-42*)[24]
- *Gunicorn* (*https://fpy.li/gunicorn*)
- *NGINX Unit* (*https://fpy.li/19-43*)

For users of the Apache HTTP server, *mod_wsgi* is the best option. It's as old as WSGI itself, but is actively maintained, and now provides a command-line launcher called `mod_wsgi-express` that makes it easier to configure and more suitable for use in Docker containers.

---

23 Some speakers spell out the WSGI acronym, while others pronounce it as one word rhyming with "whisky."

24 *uWSGI* is spelled with a lowercase "u," but that is pronounced as the Greek letter "μ," so the whole name sounds like "micro-whisky" with a "g" instead of the "k."

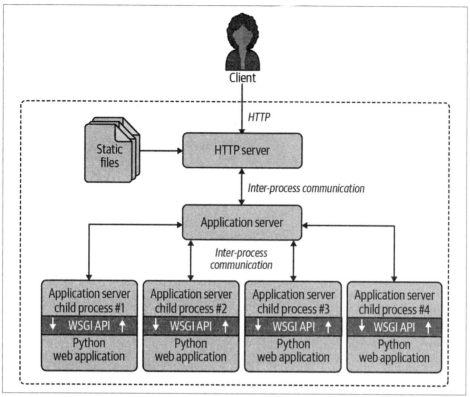

Figure 19-4. Clients connect to an HTTP server that delivers static files and routes other requests to the application server, which forks child processes to run the application code, leveraging multiple CPU cores. The WSGI API is the glue between the application server and the Python application code.

uWSGI and *Gunicorn* are the top choices in recent projects I know about. Both are often used with the *NGINX* HTTP server. *uWSGI* offers a lot of extra functionality, including an application cache, a task queue, a cron-like periodic tasks, and many other features. On the flip side, *uWSGI* is much harder to configure properly than *Gunicorn*.[25]

Released in 2018, *NGINX Unit* is a new product from the makers of the well-known *NGINX* HTTP server and reverse proxy.

---

25 Bloomberg engineers Peter Sperl and Ben Green wrote "Configuring uWSGI for Production Deployment" (*https://fpy.li/19-44*), explaining how many of the default settings in *uWSGI* are not suitable for many common deployment scenarios. Sperl presented a summary of their recommendations at EuroPython 2019 (*https://fpy.li/19-45*). Highly recommended for users of *uWSGI*.

mod_wsgi and *Gunicorn* support Python web apps only, while *uWSGI* and *NGINX Unit* work with other languages as well. Please browse their docs to learn more.

The main point: all of these application servers can potentially use all CPU cores on the server by forking multiple Python processes to run traditional web apps written in good old sequential code in *Django*, *Flask*, *Pyramid*, etc. This explains why it's been possible to earn a living as a Python web developer without ever studying the threading, multiprocessing, or asyncio modules: the application server handles concurrency transparently.

ASGI—Asynchronous Server Gateway Interface

WSGI is a synchronous API. It doesn't support coroutines with async/await—the most efficient way to implement WebSockets or HTTP long polling in Python. The ASGI specification (*https:// fpy.li/19-46*) is a successor to WSGI, designed for asynchronous Python web frameworks such as *aiohttp*, *Sanic*, *FastAPI*, etc., as well as *Django* and *Flask*, which are gradually adding asynchronous functionality.

Now let's turn to another way of bypassing the GIL to achieve higher performance with server-side Python applications.

Distributed Task Queues

When the application server delivers a request to one of the Python processes running your code, your app needs to respond quickly: you want the process to be available to handle the next request as soon as possible. However, some requests demand actions that may take longer—for example, sending email or generating a PDF. That's the problem that distributed task queues are designed to solve.

Celery (*https://fpy.li/19-47*) and *RQ* (*https://fpy.li/19-48*) are the best known open source task queues with Python APIs. Cloud providers also offer their own proprietary task queues.

These products wrap a message queue and offer a high-level API for delegating tasks to workers, possibly running on different machines.

In the context of task queues, the words *producer* and *consumer* are used instead of traditional client/server terminology. For example, a *Django* view handler *produces* job requests, which are put in the queue to be *consumed* by one or more PDF rendering processes.

Quoting directly from *Celery*'s FAQ (*https://fpy.li/19-49*), here are some typical use cases:

- Running something in the background. For example, to finish the web request as soon as possible, then update the users page incrementally. This gives the user the impression of good performance and "snappiness," even though the real work might actually take some time.

- Running something after the web request has finished.

- Making sure something is done, by executing it asynchronously and using retries.

- Scheduling periodic work.

Besides solving these immediate problems, task queues support horizontal scalability. Producers and consumers are decoupled: a producer doesn't call a consumer, it puts a request in a queue. Consumers don't need to know anything about the producers (but the request may include information about the producer, if an acknowledgment is required). Crucially, you can easily add more workers to consume tasks as demand grows. That's why *Celery* and *RQ* are called distributed task queues.

Recall that our simple *procs.py* (Example 19-13) used two queues: one for job requests, the other for collecting results. The distributed architecture of *Celery* and *RQ* uses a similar pattern. Both support using the *Redis* (*https://fpy.li/19-50*) NoSQL database as a message queue and result storage. *Celery* also supports other message queues like *RabbitMQ* or *Amazon SQS*, as well other databases for result storage.

This wraps up our introduction to concurrency in Python. The next two chapters will continue this theme, focusing on the `concurrent.futures` and `asyncio` packages of the standard library.

Chapter Summary

After a bit of theory, this chapter presented the spinner scripts implemented in each of Python's three native concurrency programming models:

- Threads, with the `threading` package

- Processes, with `multiprocessing`

- Asynchronous coroutines with `asyncio`

We then explored the real impact of the GIL with an experiment: changing the spinner examples to compute the primality of a large integer and observe the resulting behavior. This demonstrated graphically that CPU-intensive functions must be avoided in `asyncio`, as they block the event loop. The threaded version of the experiment worked—despite the GIL—because Python periodically interrupts threads, and the example used only two threads: one doing compute-intensive work, and the other driving the animation only 10 times per second. The `multiprocessing` variant worked around the GIL, starting a new process just for the animation, while the main process did the primality check.

The next example, computing several primes, highlighted the difference between mul tiprocessing and threading, proving that only processes allow Python to benefit from multicore CPUs. Python's GIL makes threads worse than sequential code for heavy computations.

The GIL dominates discussions about concurrent and parallel computing in Python, but we should not overestimate its impact. That was the point of "Python in the Multicore World" on page 729. For example, the GIL doesn't affect many use cases of Python in system administration. On the other hand, the data science and server-side development communities have worked around the GIL with industrial-strength solutions tailored to their specific needs. The last two sections mentioned two common elements to support Python server-side applications at scale: WSGI application servers and distributed task queues.

Further Reading

This chapter has an extensive reading list, so I split it into subsections.

Concurrency with Threads and Processes

The *concurrent.futures* library covered in Chapter 20 uses threads, processes, locks, and queues under the hood, but you won't see individual instances of them; they're bundled and managed by the higher-level abstractions of a ThreadPoolExecutor and a ProcessPoolExecutor. If you want to learn more about the practice of concurrent programming with those low-level objects, "An Intro to Threading in Python" (*https://fpy.li/19-51*) by Jim Anderson is a good first read. Doug Hellmann has a chapter titled "Concurrency with Processes, Threads, and Coroutines" on his website (*https://fpy.li/19-52*) and book, *The Python 3 Standard Library by Example* (*https://fpy.li/19-53*) (Addison-Wesley).

Brett Slatkin's *Effective Python* (*https://fpy.li/effectpy*), 2nd ed. (Addison-Wesley), David Beazley's *Python Essential Reference*, 4th ed. (Addison-Wesley), and Martelli et al., *Python in a Nutshell*, 3rd ed. (O'Reilly) are other general Python references with significant coverage of threading and multiprocessing. The vast multiprocessing official documentation includes useful advice in its "Programming guidelines" section (*https://fpy.li/19-54*).

Jesse Noller and Richard Oudkerk contributed the multiprocessing package, introduced in PEP 371—Addition of the multiprocessing package to the standard library (*https://fpy.li/pep371*). The official documentation for the package is a 93 KB *.rst* file (*https://fpy.li/19-55*)—that's about 63 pages—making it one of the longest chapters in the Python standard library.

In *High Performance Python*, 2nd ed., (O'Reilly), authors Micha Gorelick and Ian Ozsvald include a chapter about `multiprocessing` with an example about checking for primes with a different strategy than our *procs.py* example. For each number, they split the range of possible factors—from 2 to `sqrt(n)`—into subranges, and make each worker iterate over one of the subranges. Their divide-and-conquer approach is typical of scientific computing applications where the datasets are huge, and workstations (or clusters) have more CPU cores than users. On a server-side system handling requests from many users, it is simpler and more efficient to let each process work on one computation from start to finish—reducing the overhead of communication and coordination among processes. Besides `multiprocessing`, Gorelick and Ozsvald present many other ways of developing and deploying high-performance data science applications leveraging multiple cores, GPUs, clusters, profilers, and compilers like Cython and Numba. Their last chapter, "Lessons from the Field," is a valuable collection of short case studies contributed by other practitioners of high-performance computing in Python.

Advanced Python Development (*https://fpy.li/19-57*) by Matthew Wilkes (Apress), is a rare book that includes short examples to explain concepts, while also showing how to build a realistic application ready for production: a data aggregator, similar to DevOps monitoring systems or IoT data collectors for distributed sensors. Two chapters in *Advanced Python Development* cover concurrent programming with `thread ing` and `asyncio`.

Jan Palach's *Parallel Programming with Python* (*https://fpy.li/19-58*) (Packt, 2014) explains the core concepts behind concurrency and parallelism, covering Python's standard library as well as *Celery*.

"The Truth About Threads" is the title of Chapter 2 in *Using Asyncio in Python* by Caleb Hattingh (O'Reilly).[26] The chapter covers the benefits and drawbacks of threading—with compelling quotes from several authoritative sources—making it clear that the fundamental challenges of threads have nothing to do with Python or the GIL. Quoting verbatim from page 14 of *Using Asyncio in Python*:

> These themes repeat throughout:
>
> - Threading makes code hard to reason about.
> - Threading is an inefficient model for large-scale concurrency (thousands of concurrent tasks).

If you want to learn the hard way how difficult it is to reason about threads and locks —without risking your job—try the exercises in Allen Downey's workbook, *The Little Book of Semaphores* (*https://fpy.li/19-59*) (Green Tea Press). The exercises in

---

26 Caleb is one of the tech reviewers for this edition of *Fluent Python*.

Downey's book range from easy to very hard to unsolvable, but even the easy ones are eye-opening.

The GIL

If you are intrigued about the GIL, remember we have no control over it from Python code, so the canonical reference is in the C-API documentation: *Thread State and the Global Interpreter Lock* (*https://fpy.li/19-60*). The *Python Library and Extension FAQ* answers: *"Can't we get rid of the Global Interpreter Lock?"* (*https://fpy.li/19-61*). Also worth reading are posts by Guido van Rossum and Jesse Noller (contributor of the multiprocessing package), respectively: "It isn't Easy to Remove the GIL" (*https://fpy.li/19-62*) and "Python Threads and the Global Interpreter Lock" (*https://fpy.li/19-63*).

CPython Internals (*https://fpy.li/19-64*) by Anthony Shaw (Real Python) explains the implementation of the CPython 3 interpreter at the C programming level. Shaw's longest chapter is "Parallelism and Concurrency": a deep dive into Python's native support for threads and processes, including managing the GIL from extensions using the C/Python API.

Finally, David Beazley presented a detailed exploration in "Understanding the Python GIL" (*https://fpy.li/19-65*).[27] In slide #54 of the presentation (*https://fpy.li/19-66*), Beazley reports an increase in processing time for a particular benchmark with the new GIL algorithm introduced in Python 3.2. The issue is not significant with real workloads, according to a comment (*https://fpy.li/19-67*) by Antoine Pitrou —who implemented the new GIL algorithm—in the bug report submitted by Beazley: Python issue #7946 (*https://fpy.li/19-68*).

Concurrency Beyond the Standard Library

Fluent Python focuses on core language features and core parts of the standard library. *Full Stack Python* (*https://fpy.li/19-69*) is a great complement to this book: it's about Python's ecosystem, with sections titled "Development Environments," "Data," "Web Development," and "DevOps," among others.

I've already mentioned two books that cover concurrency using the Python standard library that also include significant content on third-party libraries and tools: *High Performance Python*, 2nd ed. and *Parallel Programming with Python*. Francesco Pierfederici's *Distributed Computing with Python* (*https://fpy.li/19-72*) (Packt) covers the standard library and also the use of cloud providers and HPC (High-Performance Computing) clusters.

---

27 Thanks to Lucas Brunialti for sending me a link to this talk.

"Python, Performance, and GPUs" (*https://fpy.li/19-73*) by Matthew Rocklin is "a status update for using GPU accelerators from Python," posted in June 2019.

"Instagram currently features the world's largest deployment of the *Django* web framework, which is written entirely in Python." That's the opening sentence of the blog post, "Web Service Efficiency at Instagram with Python" (*https://fpy.li/19-74*), written by Min Ni—a software engineer at Instagram. The post describes metrics and tools Instagram uses to optimize the efficiency of its Python codebase, as well as detect and diagnose performance regressions as it deploys its back end "30-50 times a day."

Architecture Patterns with Python: Enabling Test-Driven Development, Domain-Driven Design, and Event-Driven Microservices (*https://fpy.li/19-75*) by Harry Percival and Bob Gregory (O'Reilly) presents architectural patterns for Python server-side applications. The authors also made the book freely available online at *cosmicpython.com*.

Two elegant and easy-to-use libraries for parallelizing tasks over processes are *lelo* (*https://fpy.li/19-77*) by João S. O. Bueno and *python-parallelize* (*https://fpy.li/19-78*) by Nat Pryce. The *lelo* package defines a `@parallel` decorator that you can apply to any function to magically make it unblocking: when you call the decorated function, its execution is started in another process. Nat Pryce's *python-parallelize* package provides a `parallelize` generator that distributes the execution of a `for` loop over multiple CPUs. Both packages are built on the *multiprocessing* library.

Python core developer Eric Snow maintains a Multicore Python (*https://fpy.li/19-79*) wiki, with notes about his and other people's efforts to improve Python's support for parallel execution. Snow is the author of PEP 554—Multiple Interpreters in the Stdlib (*https://fpy.li/pep554*). If approved and implemented, PEP 554 lays the groundwork for future enhancements that may eventually allow Python to use multiple cores without the overheads of *multiprocessing*. One of the biggest blockers is the complex interaction between multiple active subinterpreters and extensions that assume a single interpreter.

Mark Shannon—also a Python maintainer—created a useful table (*https://fpy.li/19-80*) comparing concurrent models in Python, referenced in a discussion about subinterpreters between him, Eric Snow, and other developers on the python-dev (*https://fpy.li/19-81*) mailing list. In Shannon's table, the "Ideal CSP" column refers to the theoretical Communicating sequential processes (*https://fpy.li/19-82*) model proposed by Tony Hoare in 1978. Go also allows shared objects, violating an essential constraint of CSP: execution units should communicate through message passing through channels.

Stackless Python (*https://fpy.li/19-83*) (a.k.a. *Stackless*) is a fork of CPython implementing microthreads, which are application-level lightweight threads—as opposed to OS threads. The massively multiplayer online game *EVE Online* (*https://fpy.li/19-84*) was built on *Stackless*, and engineers employed by the game company CCP (*https://fpy.li/19-85*) were maintainers of *Stackless* (*https://fpy.li/19-86*) for a while. Some features of *Stackless* were reimplemented in the *Pypy* (*https://fpy.li/19-87*) interpreter and the *greenlet* (*https://fpy.li/19-14*) package, the core technology of the *gevent* (*https://fpy.li/19-17*) networking library, which in turn is the foundation of the *Gunicorn* (*https://fpy.li/gunicorn*) application server.

The actor model of concurrent programming is at the core of the highly scalable Erlang and Elixir languages, and is also the model of the Akka framework for Scala and Java. If you want to try out the actor model in Python, check out the *Thespian* (*https://fpy.li/19-90*) and *Pykka* (*https://fpy.li/19-91*) libraries.

My remaining recommendations have few or zero mentions of Python, but are nevertheless relevant to readers interested in the theme of this chapter.

Concurrency and Scalability Beyond Python

RabbitMQ in Action (*https://fpy.li/19-92*) by Alvaro Videla and Jason J. W. Williams (Manning) is a very well-written introduction to *RabbitMQ* and the Advanced Message Queuing Protocol (AMQP) standard, with examples in Python, PHP, and Ruby. Regardless of the rest of your tech stack, and even if you plan to use *Celery* with *RabbitMQ* under the hood, I recommend this book for its coverage of concepts, motivation, and patterns for distributed message queues, as well as operating and tuning *RabbitMQ* at scale.

I learned a lot reading *Seven Concurrency Models in Seven Weeks* (*https://fpy.li/19-93*), by Paul Butcher (Pragmatic Bookshelf), with the eloquent subtitle *When Threads Unravel*. Chapter 1 of the book presents the core concepts and challenges of programming with threads and locks in Java.[28] The remaining six chapters of the book are devoted to what the author considers better alternatives for concurrent and parallel programming, as supported by different languages, tools, and libraries. The examples use Java, Clojure, Elixir, and C (for the chapter about parallel programming with the OpenCL framework (*https://fpy.li/19-94*)). The CSP model is exemplified with Clojure code, although the Go language deserves credit for popularizing that approach. Elixir is the language of the examples illustrating the actor model. A freely available, alternative bonus chapter (*https://fpy.li/19-95*) about actors uses Scala and the Akka framework. Unless you already know Scala, Elixir is a more accessible language to learn and experiment with the actor model and the Erlang/OTP distributed systems platform.

---

28 Python's `threading` and `concurrent.futures` APIs are heavily influenced by the Java standard library.

Unmesh Joshi of Thoughtworks has contributed several pages documenting "Patterns of Distributed Systems" to Martin Fowler's blog (*https://fpy.li/19-96*). The opening page (*https://fpy.li/19-97*) is a great introduction the topic, with links to individual patterns. Joshi is adding patterns incrementally, but what's already there distills years of hard-earned experience in mission-critical systems.

Martin Kleppmann's *Designing Data-Intensive Applications* (O'Reilly) is a rare book written by a practitioner with deep industry experience and advanced academic background. The author worked with large-scale data infrastructure at LinkedIn and two startups, before becoming a researcher of distributed systems at the University of Cambridge. Each chapter in Kleppmann's book ends with an extensive list of references, including recent research results. The book also includes numerous illuminating diagrams and beautiful concept maps.

I was fortunate to be in the audience for Francesco Cesarini's outstanding workshop on the architecture of reliable distributed systems at OSCON 2016: "Designing and architecting for scalability with Erlang/OTP" (video (*https://fpy.li/19-99*) at the O'Reilly Learning Platform). Despite the title, 9:35 into the video, Cesarini explains:

> Very little of what I am going to say will be Erlang-specific [...]. The fact remains that Erlang will remove a lot of accidental difficulties to making systems which are resilient and which never fail, and are scalable. So it will be much easier if you do use Erlang, or a language running on the Erlang virtual machine.

That workshop was based on the last four chapters of *Designing for Scalability with Erlang/OTP* (*https://fpy.li/19-100*) by Francesco Cesarini and Steve Vinoski (O'Reilly).

Programming distributed systems is challenging and exciting, but beware of *web-scale envy* (*https://fpy.li/19-40*). The KISS principle (*https://fpy.li/19-102*) remains solid engineering advice.

Check out the paper "Scalability! But at what COST?" (*https://fpy.li/19-103*) by Frank McSherry, Michael Isard, and Derek G. Murray. The authors identified parallel graph-processing systems presented in academic symposia that require hundreds of cores to outperform a "competent single-threaded implementation." They also found systems that "underperform one thread for all of their reported configurations."

Those findings remind me of a classic hacker quip:

> My Perl script is faster than your Hadoop cluster.

Soapbox

To Manage Complexity, We Need Constraints

I learned to program on a TI-58 calculator. Its "language" was similar to assembly. At that level, all "variables" are globals, and you don't have the luxury of structured control flow statements. You have conditional jumps: instructions that take the execution directly to an arbitrary location—ahead or behind the current spot—depending on the value of a CPU register or flag.

Basically you can do anything in assembly, and that's the challenge: there are very few constraints to keep you from making mistakes, and to help maintainers understand the code when changes are needed.

The second language I learned was the unstructured BASIC that came with 8-bit computers—nothing like Visual Basic, which appeared much later. There were FOR, GOSUB, and RETURN statements, but still no concept of local variables. GOSUB did not support parameter passing: it was just a fancy GOTO that put a return line number in a stack so that RETURN had a target to jump to. Subroutines could help themselves to the global data, and put results there too. We had to improvise other forms of control flow with combinations of IF and GOTO—which, again, allowed you to jump to any line of the program.

After a few years of programming with jumps and global variables, I remember the struggle to rewire my brain for "structured programming" when I learned Pascal. Now I had to use control flow statements around blocks of code that have a single entry point. I couldn't jump to any instruction I liked. Global variables were unavoidable in BASIC, but now they were taboo. I needed to rethink the flow of data and explicitly pass arguments to functions.

The next challenge for me was learning object-oriented programming. At its core, object-oriented programming is structured programming with more constraints and polymorphism. Information hiding forces yet another rethink of where data lives. I remember being frustrated more than once because I had to refactor my code so that a method I was writing could get information that was encapsulated in an object that my method could not reach.

Functional programming languages add other constraints, but immutability is the hardest to swallow after decades of imperative programming and object-oriented programming. After we get used to these constraints, we see them as blessings. They make reasoning about the code much easier.

Lack of constraints is the main problem with the threads-and-locks model of concurrent programming. When summarizing Chapter 1 of *Seven Concurrency Models in Seven Weeks*, Paul Butcher wrote:

The greatest weakness of the approach, however, is that threads-and-locks programming is *hard*. It may be easy for a language designer to add them to a language, but they provide us, the poor programmers, with very little help.

Some examples of unconstrained behavior in that model:

- Threads can share access to arbitrary, mutable data structures.
- The scheduler can interrupt a thread at almost any point, including in the middle of a simple operation like a += 1. Very few operations are atomic at the level of source code expressions.
- Locks are usually *advisory*. That's a technical term meaning that you must remember to explicitly hold a lock before updating a shared data structure. If you forget to get the lock, nothing prevents your code from messing up the data while another thread dutifully holds the lock and is updating the same data.

In contrast, consider some constraints enforced by the actor model, in which the execution unit is called an *actor*:[29]

- An actor can have internal state, but cannot share state with other actors.
- Actors can only communicate by sending and receiving messages.
- Messages only hold copies of data, not references to mutable data.
- An actor only handles one message at a time. There is no concurrent execution inside a single actor.

Of course, you can adopt an *actor style* of coding in any language by following these rules. You can also use object-oriented programming idioms in C, and even structured programming patterns in assembly. But doing any of that requires a lot of agreement and discipline among everyone who touches the code.

Managing locks is unnecessary in the actor model, as implemented by Erlang and Elixir, where all data types are immutable.

Threads-and-locks are not going away. I just don't think dealing with such low-level entities is a good use of my time as I write applications—as opposed to kernel modules or databases.

I reserve the right to change my mind, always. But right now, I am convinced that the actor model is the most sensible, general-purpose concurrent programming model available. CSP (Communicating Sequential Processes) is also sensible, but its

---

29 The Erlang community uses the term "process" for actors. In Erlang, each process is a function in its own loop, so they are very lightweight and it's feasible to have millions of them active at once in a single machine —no relation to the heavyweight OS processes we've been talking about elsewhere in this chapter. So here we have examples of the two sins described by Prof. Simon: using different words to mean the same thing, and using one word to mean different things.

implementation in Go leaves out some constraints. The idea in CSP is that coroutines (or *goroutines* in Go) exchange data and synchronize using queues (called *channels* in Go). But Go also supports memory sharing and locks. I've seen a book about Go advocate the use of shared memory and locks instead of channels—in the name of performance. Old habits die hard.

## CHAPTER 20

Concurrent Executors

The people bashing threads are typically system programmers which have in mind use cases that the typical application programmer will never encounter in her life. [...] In 99% of the use cases an application programmer is likely to run into, the simple pattern of spawning a bunch of independent threads and collecting the results in a queue is everything one needs to know.

—Michele Simionato, Python deep thinker[1]

This chapter focuses on the `concurrent.futures.Executor` classes that encapsulate the pattern of "spawning a bunch of independent threads and collecting the results in a queue," described by Michele Simionato. The concurrent executors make this pattern almost trivial to use, not only with threads but also with processes—useful for compute-intensive tasks.

Here I also introduce the concept of *futures*—objects representing the asynchronous execution of an operation, similar to JavaScript promises. This primitive idea is the foundation not only of `concurrent.futures` but also of the `asyncio` package, the subject of Chapter 21.

What's New in This Chapter

I renamed the chapter from "Concurrency with Futures" to "Concurrent Executors" because the executors are the most important high-level feature covered here. Futures are low-level objects, focused on in "Where Are the Futures?" on page 755, but mostly invisible in the rest of the chapter.

---

1 From Michele Simionato's post, "Threads, processes and concurrency in Python: some thoughts" (*https:// fpy.li/20-1*), summarized as "Removing the hype around the multicore (non) revolution and some (hopefully) sensible comment about threads and other forms of concurrency."

All the HTTP client examples now use the new *HTTPX* (*https://fpy.li/httpx*) library, which provides synchronous and asynchronous APIs.

The setup for the experiments in "Downloads with Progress Display and Error Handling" on page 766 is now simpler, thanks to the multithreaded server added to the http.server (*https://fpy.li/20-2*) package in Python 3.7. Previously, the standard library only had the single-threaded BaseHttpServer, which was no good for experimenting with concurrent clients, so I had to resort to external tools in the first edition of this book.

"Launching Processes with concurrent.futures" on page 758 now demonstrates how an executor simplifies the code we saw in "Code for the Multicore Prime Checker" on page 723.

Finally, I moved most of the theory to the new Chapter 19, "Concurrency Models in Python".

Concurrent Web Downloads

Concurrency is essential for efficient network I/O: instead of idly waiting for remote machines, the application should do something else until a response comes back.[2]

To demonstrate with code, I wrote three simple programs to download images of 20 country flags from the web. The first one, *flags.py*, runs sequentially: it only requests the next image when the previous one is downloaded and saved locally. The other two scripts make concurrent downloads: they request several images practically at the same time, and save them as they arrive. The *flags_threadpool.py* script uses the con current.futures package, while *flags_asyncio.py* uses asyncio.

Example 20-1 shows the result of running the three scripts, three times each. I also posted a 73s video on YouTube (*https://fpy.li/20-3*) so you can watch them running while a macOS Finder window displays the flags as they are saved. The scripts are downloading images from *fluentpython.com*, which is behind a CDN, so you may see slower results in the first runs. The results in Example 20-1 were obtained after several runs, so the CDN cache was warm.

Example 20-1. Three typical runs of the scripts flags.py, flags_threadpool.py, and flags_asyncio.py

```
$ python3 flags.py
BD BR CD CN DE EG ET FR ID IN IR JP MX NG PH PK RU TR US VN  ❶
20 flags downloaded in 7.26s  ❷
$ python3 flags.py
```

---

2 Particularly if your cloud provider rents machines by the second, regardless of how busy the CPUs are.

```
BD BR CD CN DE EG ET FR ID IN IR JP MX NG PH PK RU TR US VN
20 flags downloaded in 7.20s
$ python3 flags.py
BD BR CD CN DE EG ET FR ID IN IR JP MX NG PH PK RU TR US VN
20 flags downloaded in 7.09s
$ python3 flags_threadpool.py
DE BD CN JP ID EG NG BR RU CD IR MX US PH FR PK VN IN ET TR
20 flags downloaded in 1.37s  ❸
$ python3 flags_threadpool.py
EG BR FR IN BD JP DE RU PK PH CD MX ID US NG TR CN VN ET IR
20 flags downloaded in 1.60s
$ python3 flags_threadpool.py
BD DE EG CN ID RU IN VN ET MX FR CD NG US JP TR PK BR IR PH
20 flags downloaded in 1.22s
$ python3 flags_asyncio.py  ❹
BD BR IN ID TR DE CN US IR PK PH FR RU NG VN ET MX EG JP CD
20 flags downloaded in 1.36s
$ python3 flags_asyncio.py
RU CN BR IN FR BD TR EG VN IR PH CD ET ID NG DE JP PK MX US
20 flags downloaded in 1.27s
$ python3 flags_asyncio.py
RU IN ID DE BR VN PK MX US IR ET EG NG BD FR CN JP PH CD TR  ❺
20 flags downloaded in 1.42s
```

❶ The output for each run starts with the country codes of the flags as they are downloaded, and ends with a message stating the elapsed time.

❷ It took *flags.py* an average 7.18s to download 20 images.

❸ The average for *flags_threadpool.py* was 1.40s.

❹ For *flags_asyncio.py*, 1.35s was the average time.

❺ Note the order of the country codes: the downloads happened in a different order every time with the concurrent scripts.

The difference in performance between the concurrent scripts is not significant, but they are both more than five times faster than the sequential script—and this is just for the small task of downloading 20 files of a few kilobytes each. If you scale the task to hundreds of downloads, the concurrent scripts can outpace the sequential code by a factor or 20 or more.

> While testing concurrent HTTP clients against public web servers, you may inadvertently launch a denial-of-service (DoS) attack, or be suspected of doing so. In the case of Example 20-1, it's OK to do it because those scripts are hardcoded to make only 20 requests. We'll use Python's `http.server` package to run tests later in this chapter.

Now let's study the implementations of two of the scripts tested in Example 20-1: *flags.py* and *flags_threadpool.py*. I will leave the third script, *flags_asyncio.py*, for Chapter 21, but I wanted to demonstrate all three together to make two points:

1. Regardless of the concurrency constructs you use—threads or coroutines—you'll see vastly improved throughput over sequential code in network I/O operations, if you code it properly.

2. For HTTP clients that can control how many requests they make, there is no significant difference in performance between threads and coroutines.[3]

On to the code.

A Sequential Download Script

Example 20-2 contains the implementation of *flags.py*, the first script we ran in Example 20-1. It's not very interesting, but we'll reuse most of its code and settings to implement the concurrent scripts, so it deserves some attention.

> For clarity, there is no error handling in Example 20-2. We will deal with exceptions later, but here I want to focus on the basic structure of the code, to make it easier to contrast this script with the concurrent ones.

Example 20-2. flags.py: sequential download script; some functions will be reused by the other scripts

```python
import time
from pathlib import Path
from typing import Callable

import httpx                                            ❶

POP20_CC = ('CN IN US ID BR PK NG BD RU JP '
            'MX PH VN ET EG DE IR TR CD FR').split()    ❷

BASE_URL = 'https://www.fluentpython.com/data/flags'    ❸
DEST_DIR = Path('downloaded')                           ❹

def save_flag(img: bytes, filename: str) -> None:       ❺
    (DEST_DIR / filename).write_bytes(img)
```

---

3 For servers that may be hit by many clients, there is a difference: coroutines scale better because they use much less memory than threads, and also reduce the cost of context switching, which I mentioned in "Thread-Based Nonsolution" on page 728.

```
def get_flag(cc: str) -> bytes:    ❻
    url = f'{BASE_URL}/{cc}/{cc}.gif'.lower()
    resp = httpx.get(url, timeout=6.1,      ❼
                     follow_redirects=True)  ❽
    resp.raise_for_status()  ❾
    return resp.content

def download_many(cc_list: list[str]) -> int:  ❿
    for cc in sorted(cc_list):                 ⓫
        image = get_flag(cc)
        save_flag(image, f'{cc}.gif')
        print(cc, end=' ', flush=True)         ⓬
    return len(cc_list)

def main(downloader: Callable[[list[str]], int]) -> None:  ⓭
    DEST_DIR.mkdir(exist_ok=True)                          ⓮
    t0 = time.perf_counter()                               ⓯
    count = downloader(POP20_CC)
    elapsed = time.perf_counter() - t0
    print(f'\n{count} downloads in {elapsed:.2f}s')

if __name__ == '__main__':
    main(download_many)      ⓰
```

❶ Import the httpx library. It's not part of the standard library, so by convention the import goes after the standard library modules and a blank line.

❷ List of the ISO 3166 country codes for the 20 most populous countries in order of decreasing population.

❸ The directory with the flag images.[4]

❹ Local directory where the images are saved.

❺ Save the img bytes to filename in the DEST_DIR.

❻ Given a country code, build the URL and download the image, returning the binary contents of the response.

❼ It's good practice to add a sensible timeout to network operations, to avoid blocking for several minutes for no good reason.

---

4 The images are originally from the CIA World Factbook (*https://fpy.li/20-4*), a public-domain, US government publication. I copied them to my site to avoid the risk of launching a DOS attack on *cia.gov*.

❽ By default, *HTTPX* does not follow redirects.[5]

❾ There's no error handling in this script, but this method raises an exception if the HTTP status is not in the 2XX range—highly recommended to avoid silent failures.

❿ `download_many` is the key function to compare with the concurrent implementations.

⓫ Loop over the list of country codes in alphabetical order, to make it easy to see that the ordering is preserved in the output; return the number of country codes downloaded.

⓬ Display one country code at a time in the same line so we can see progress as each download happens. The `end=' '` argument replaces the usual line break at the end of each line printed with a space character, so all country codes are displayed progressively in the same line. The `flush=True` argument is needed because, by default, Python output is line buffered, meaning that Python only displays printed characters after a line break.

⓭ `main` must be called with the function that will make the downloads; that way, we can use `main` as a library function with other implementations of `download_many` in the `threadpool` and `ascyncio` examples.

⓮ Create `DEST_DIR` if needed; don't raise an error if the directory exists.

⓯ Record and report the elapsed time after running the `downloader` function.

⓰ Call `main` with the `download_many` function.

 The *HTTPX* (*https://fpy.li/httpx*) library is inspired by the Pythonic *requests* (*https://fpy.li/20-5*) package, but is built on a more modern foundation. Crucially, *HTTPX* provides synchronous and asynchronous APIs, so we can use it in all HTTP client examples in this chapter and the next. Python's standard library provides the `url lib.request` module, but its API is synchronous only, and is not user friendly.

---

5 Setting `follow_redirects=True` is not needed for this example, but I wanted to highlight this important difference between *HTTPX* and *requests*. Also, setting `follow_redirects=True` in this example gives me flexibility to host the image files elsewhere in the future. I think the *HTTPX* default setting of `follow_redirects =False` is sensible because unexpected redirects can mask needless requests and complicate error diagnostics.

There's really nothing new to *flags.py*. It serves as a baseline for comparing the other scripts, and I used it as a library to avoid redundant code when implementing them. Now let's see a reimplementation using concurrent.futures.

Downloading with concurrent.futures

The main features of the concurrent.futures package are the ThreadPoolExecutor and ProcessPoolExecutor classes, which implement an API to submit callables for execution in different threads or processes, respectively. The classes transparently manage a pool of worker threads or processes, and queues to distribute jobs and collect results. But the interface is very high-level, and we don't need to know about any of those details for a simple use case like our flag downloads.

Example 20-3 shows the easiest way to implement the downloads concurrently, using the ThreadPoolExecutor.map method.

Example 20-3. flags_threadpool.py: threaded download script using futures.Thread PoolExecutor

```
from concurrent import futures

from flags import save_flag, get_flag, main      ❶

def download_one(cc: str):      ❷
    image = get_flag(cc)
    save_flag(image, f'{cc}.gif')
    print(cc, end=' ', flush=True)
    return cc

def download_many(cc_list: list[str]) -> int:
    with futures.ThreadPoolExecutor() as executor:      ❸
        res = executor.map(download_one, sorted(cc_list))      ❹

    return len(list(res))      ❺

if __name__ == '__main__':
    main(download_many)      ❻
```

❶ Reuse some functions from the flags module (Example 20-2).

❷ Function to download a single image; this is what each worker will execute.

❸ Instantiate the ThreadPoolExecutor as a context manager; the executor .__exit__ method will call executor.shutdown(wait=True), which will block until all threads are done.

❹ The map method is similar to the map built-in, except that the download_one function will be called concurrently from multiple threads; it returns a generator that you can iterate to retrieve the value returned by each function call—in this case, each call to download_one will return a country code.

❺ Return the number of results obtained. If any of the threaded calls raises an exception, that exception is raised here when the implicit next() call inside the list constructor tries to retrieve the corresponding return value from the iterator returned by executor.map.

❻ Call the main function from the flags module, passing the concurrent version of download_many.

Note that the download_one function from Example 20-3 is essentially the body of the for loop in the download_many function from Example 20-2. This is a common refactoring when writing concurrent code: turning the body of a sequential for loop into a function to be called concurrently.

 Example 20-3 is very short because I was able to reuse most functions from the sequential *flags.py* script. One of the best features of concurrent.futures is to make it simple to add concurrent execution on top of legacy sequential code.

The ThreadPoolExecutor constructor takes several arguments not shown, but the first and most important one is max_workers, setting the maximum number of worker threads to be executed. When max_workers is None (the default), ThreadPool Executor decides its value using the following expression—since Python 3.8:

```
max_workers = min(32, os.cpu_count() + 4)
```

The rationale is explained in the ThreadPoolExecutor documentation (*https://fpy.li/ 20-6*):

> This default value preserves at least 5 workers for I/O bound tasks. It utilizes at most 32 CPU cores for CPU bound tasks which release the GIL. And it avoids using very large resources implicitly on many-core machines.
>
> ThreadPoolExecutor now reuses idle worker threads before starting max_workers worker threads too.

To conclude: the computed default for max_workers is sensible, and ThreadPoolExe cutor avoids starting new workers unnecessarily. Understanding the logic behind max_workers may help you decide when and how to set it yourself.

The library is called *concurrency.futures*, yet there are no futures to be seen in Example 20-3, so you may be wondering where they are. The next section explains.

Where Are the Futures?

Futures are core components of `concurrent.futures` and of `asyncio`, but as users of these libraries we sometimes don't see them. Example 20-3 depends on futures behind the scenes, but the code I wrote does not touch them directly. This section is an overview of futures, with an example that shows them in action.

Since Python 3.4, there are two classes named `Future` in the standard library: `concurrent.futures.Future` and `asyncio.Future`. They serve the same purpose: an instance of either `Future` class represents a deferred computation that may or may not have completed. This is somewhat similar to the `Deferred` class in Twisted, the `Future` class in Tornado, and `Promise` in modern JavaScript.

Futures encapsulate pending operations so that we can put them in queues, check whether they are done, and retrieve results (or exceptions) when they become available.

An important thing to know about futures is that you and I should not create them: they are meant to be instantiated exclusively by the concurrency framework, be it `concurrent.futures` or `asyncio`. Here is why: a `Future` represents something that will eventually run, therefore it must be scheduled to run, and that's the job of the framework. In particular, `concurrent.futures.Future` instances are created only as the result of submitting a callable for execution with a `concurrent.futures.Executor` subclass. For example, the `Executor.submit()` method takes a callable, schedules it to run, and returns a `Future`.

Application code is not supposed to change the state of a future: the concurrency framework changes the state of a future when the computation it represents is done, and we can't control when that happens.

Both types of `Future` have a `.done()` method that is nonblocking and returns a Boolean that tells you whether the callable wrapped by that future has executed or not. However, instead of repeatedly asking whether a future is done, client code usually asks to be notified. That's why both `Future` classes have an `.add_done_callback()` method: you give it a callable, and the callable will be invoked with the future as the single argument when the future is done. Be aware that the callback callable will run in the same worker thread or process that ran the function wrapped in the future.

There is also a `.result()` method, which works the same in both classes when the future is done: it returns the result of the callable, or re-raises whatever exception might have been thrown when the callable was executed. However, when the future is not done, the behavior of the result method is very different between the two flavors of Future. In a concurrency.futures.Future instance, invoking f.result() will block the caller's thread until the result is ready. An optional timeout argument can be passed, and if the future is not done in the specified time, the result method raises TimeoutError. The asyncio.Future.result method does not support time-out, and await is the preferred way to get the result of futures in asyncio—but await doesn't work with concurrency.futures.Future instances.

Several functions in both libraries return futures; others use them in their implementation in a way that is transparent to the user. An example of the latter is the Executor.map we saw in Example 20-3: it returns an iterator in which __next__ calls the result method of each future, so we get the results of the futures, and not the futures themselves.

To get a practical look at futures, we can rewrite Example 20-3 to use the concurrent.futures.as_completed (*https://fpy.li/20-7*) function, which takes an iterable of futures and returns an iterator that yields futures as they are done.

Using futures.as_completed requires changes to the download_many function only. The higher-level executor.map call is replaced by two for loops: one to create and schedule the futures, the other to retrieve their results. While we are at it, we'll add a few print calls to display each future before and after it's done. Example 20-4 shows the code for a new download_many function. The code for download_many grew from 5 to 17 lines, but now we get to inspect the mysterious futures. The remaining functions are the same as in Example 20-3.

Example 20-4. flags_threadpool_futures.py: replacing executor.map with executor.submit and futures.as_completed in the download_many function

```python
def download_many(cc_list: list[str]) -> int:
    cc_list = cc_list[:5]  ❶
    with futures.ThreadPoolExecutor(max_workers=3) as executor:  ❷
        to_do: list[futures.Future] = []
        for cc in sorted(cc_list):  ❸
            future = executor.submit(download_one, cc)  ❹
            to_do.append(future)  ❺
            print(f'Scheduled for {cc}: {future}')  ❻

        for count, future in enumerate(futures.as_completed(to_do), 1):  ❼
            res: str = future.result()  ❽
            print(f'{future} result: {res!r}')  ❾

    return count
```

❶ For this demonstration, use only the top five most populous countries.

❷ Set max_workers to 3 so we can see pending futures in the output.

❸ Iterate over country codes alphabetically, to make it clear that results will arrive out of order.

❹ executor.submit schedules the callable to be executed, and returns a future representing this pending operation.

❺ Store each future so we can later retrieve them with as_completed.

❻ Display a message with the country code and the respective future.

❼ as_completed yields futures as they are completed.

❽ Get the result of this future.

❾ Display the future and its result.

Note that the future.result() call will never block in this example because the future is coming out of as_completed. Example 20-5 shows the output of one run of Example 20-4.

Example 20-5. Output of flags_threadpool_futures.py

```
$ python3 flags_threadpool_futures.py
Scheduled for BR: <Future at 0x100791518 state=running>   ❶
Scheduled for CN: <Future at 0x100791710 state=running>
Scheduled for ID: <Future at 0x100791a90 state=running>
Scheduled for IN: <Future at 0x101807080 state=pending>   ❷
Scheduled for US: <Future at 0x101807128 state=pending>
CN <Future at 0x100791710 state=finished returned str> result: 'CN'   ❸
BR ID <Future at 0x100791518 state=finished returned str> result: 'BR'   ❹
<Future at 0x100791a90 state=finished returned str> result: 'ID'
IN <Future at 0x101807080 state=finished returned str> result: 'IN'
US <Future at 0x101807128 state=finished returned str> result: 'US'

5 downloads in 0.70s
```

❶ The futures are scheduled in alphabetical order; the repr() of a future shows its state: the first three are running, because there are three worker threads.

❷ The last two futures are pending, waiting for worker threads.

❸ The first CN here is the output of download_one in a worker thread; the rest of the line is the output of download_many.

❹ Here, two threads output codes before download_many in the main thread can display the result of the first thread.

 I recommend experimenting with *flags_threadpool_futures.py*. If you run it several times, you'll see the order of the results varying. Increasing max_workers to 5 will increase the variation in the order of the results. Decreasing it to 1 will make this script run sequentially, and the order of the results will always be the order of the submit calls.

We saw two variants of the download script using concurrent.futures: one in Example 20-3 with ThreadPoolExecutor.map and one in Example 20-4 with futures.as_completed. If you are curious about the code for *flags_asyncio.py*, you may peek at Example 21-3 in Chapter 21, where it is explained.

Now let's take a brief look at a simple way to work around the GIL for CPU-bound jobs using concurrent.futures.

Launching Processes with concurrent.futures

The concurrent.futures documentation page (*https://fpy.li/20-8*) is subtitled "Launching parallel tasks." The package enables parallel computation on multicore machines because it supports distributing work among multiple Python processes using the ProcessPoolExecutor class.

Both ProcessPoolExecutor and ThreadPoolExecutor implement the Executor (*https://fpy.li/20-9*) interface, so it's easy to switch from a thread-based to a process-based solution using concurrent.futures.

There is no advantage in using a ProcessPoolExecutor for the flags download example or any I/O-bound job. It's easy to verify this; just change these lines in Example 20-3:

```
def download_many(cc_list: list[str]) -> int:
    with futures.ThreadPoolExecutor() as executor:
```

To this:

```
def download_many(cc_list: list[str]) -> int:
    with futures.ProcessPoolExecutor() as executor:
```

The constructor for ProcessPoolExecutor also has a max_workers parameter, which defaults to None. In that case, the executor limits the number of workers to the number returned by os.cpu_count().

Processes use more memory and take longer to start than threads, so the real value of ProcessPoolExecutor is in CPU-intensive jobs. Let's go back to the primality test example of "A Homegrown Process Pool" on page 720, rewriting it with concurrent.futures.

Multicore Prime Checker Redux

In "Code for the Multicore Prime Checker" on page 723 we studied *procs.py*, a script that checked the primality of some large numbers using multiprocessing. In Example 20-6 we solve the same problem in the *proc_pool.py* program using a Proc essPoolExecutor. From the first import to the main() call at the end, *procs.py* has 43 nonblank lines of code, and *proc_pool.py* has 31—28% shorter.

Example 20-6. proc_pool.py: procs.py rewritten with ProcessPoolExecutor

```
import sys
from concurrent import futures     ❶
from time import perf_counter
from typing import NamedTuple

from primes import is_prime, NUMBERS

class PrimeResult(NamedTuple):     ❷
    n: int
    flag: bool
    elapsed: float

def check(n: int) -> PrimeResult:
    t0 = perf_counter()
    res = is_prime(n)
    return PrimeResult(n, res, perf_counter() - t0)

def main() -> None:
    if len(sys.argv) < 2:
        workers = None        ❸
    else:
        workers = int(sys.argv[1])

    executor = futures.ProcessPoolExecutor(workers)     ❹
    actual_workers = executor._max_workers  # type: ignore     ❺

    print(f'Checking {len(NUMBERS)} numbers with {actual_workers} processes:')

    t0 = perf_counter()
```

```
        numbers = sorted(NUMBERS, reverse=True)  ❻
        with executor:  ❼
            for n, prime, elapsed in executor.map(check, numbers):  ❽
                label = 'P' if prime else ' '
                print(f'{n:16}  {label} {elapsed:9.6f}s')

        time = perf_counter() - t0
        print(f'Total time: {time:.2f}s')

if __name__ == '__main__':
    main()
```

❶ No need to import multiprocessing, SimpleQueue etc.; concurrent.futures hides all that.

❷ The PrimeResult tuple and the check function are the same as we saw in *procs.py*, but we don't need the queues and the worker function anymore.

❸ Instead of deciding ourselves how many workers to use if no command-line argument was given, we set workers to None and let the ProcessPoolExecutor decide.

❹ Here I build the ProcessPoolExecutor before the with block in ❼ so that I can display the actual number of workers in the next line.

❺ _max_workers is an undocumented instance attribute of a ProcessPoolExecutor. I decided to use it to show the number of workers when the workers variable is None. *Mypy* correctly complains when I access it, so I put the type: ignore comment to silence it.

❻ Sort the numbers to be checked in descending order. This will expose a difference in the behavior of *proc_pool.py* when compared with *procs.py*. See the explanation after this example.

❼ Use the executor as a context manager.

❽ The executor.map call returns the PrimeResult instances returned by check in the same order as the numbers arguments.

If you run Example 20-6, you'll see the results appearing in strict descending order, as shown in Example 20-7. In contrast, the ordering of the output of *procs.py* (shown in "Process-Based Solution" on page 722) is heavily influenced by the difficulty in checking whether each number is a prime. For example, *procs.py* shows the result for

7777777777777777 near the top, because it has a low divisor, 7, so is_prime quickly determines it's not a prime.

In contrast, 7777777536340681 is 88191709^2, so is_prime will take much longer to determine that it's a composite number, and even longer to find out that 7777777777777753 is prime—therefore both of these numbers appear near the end of the output of *procs.py*.

Running *proc_pool.py*, you'll observe not only the descending order of the results, but also that the program will appear to be stuck after showing the result for 9999999999999999.

Example 20-7. Output of proc_pool.py

```
$ ./proc_pool.py
Checking 20 numbers with 12 processes:
9999999999999999      0.000024s  ❶
9999999999999917  P   9.500677s  ❷
7777777777777777      0.000022s  ❸
7777777777777753  P   8.976933s
7777777536340681      8.896149s
6666667141414921      8.537621s
6666666666666719  P   8.548641s
6666666666666666      0.000002s
5555555555555555      0.000017s
5555555555555503  P   8.214086s
5555553133149889      8.067247s
4444444488888889      7.546234s
4444444444444444      0.000002s
4444444444444423  P   7.622370s
3333335652092209      6.724649s
3333333333333333      0.000018s
3333333333333301  P   6.655039s
 295593572317531  P   2.072723s
 142702110479723  P   1.461840s
               2  P   0.000001s
Total time: 9.65s
```

❶ This line appears very quickly.

❷ This line takes more than 9.5s to show up.

❸ All the remaining lines appear almost immediately.

Here is why *proc_pool.py* behaves in that way:

- As mentioned before, executor.map(check, numbers) returns the result in the same order as the numbers are given.
- By default, *proc_pool.py* uses as many workers as there are CPUs—it's what ProcessPoolExecutor does when max_workers is None. That's 12 processes in this laptop.
- Because we are submitting numbers in descending order, the first is 9999999999999999; with 9 as a divisor, it returns quickly.
- The second number is 9999999999999917, the largest prime in the sample. This will take longer than all the others to check.
- Meanwhile, the remaining 11 processes will be checking other numbers, which are either primes or composites with large factors, or composites with very small factors.
- When the worker in charge of 9999999999999917 finally determines that's a prime, all the other processes have completed their last jobs, so the results appear immediately after.

Although the progress of *proc_pool.py* is not as visible as that of *procs.py*, the overall execution time is practically the same as depicted in Figure 19-2, for the same number of workers and CPU cores.

Understanding how concurrent programs behave is not straightforward, so here's is a second experiment that may help you visualize the operation of Executor.map.

Experimenting with Executor.map

Let's investigate Executor.map, now using a ThreadPoolExecutor with three workers running five callables that output timestamped messages. The code is in Example 20-8, the output in Example 20-9.

Example 20-8. demo_executor_map.py: Simple demonstration of the map method of ThreadPoolExecutor

```
from time import sleep, strftime
from concurrent import futures

def display(*args):      ❶
    print(strftime('[%H:%M:%S]'), end=' ')
    print(*args)
```

```
def loiter(n):  ❷
    msg = '{}loiter({}): doing nothing for {}s...'
    display(msg.format('\t'*n, n, n))
    sleep(n)
    msg = '{}loiter({}): done.'
    display(msg.format('\t'*n, n))
    return n * 10  ❸

def main():
    display('Script starting.')
    executor = futures.ThreadPoolExecutor(max_workers=3)  ❹
    results = executor.map(loiter, range(5))  ❺
    display('results:', results)  ❻
    display('Waiting for individual results:')
    for i, result in enumerate(results):  ❼
        display(f'result {i}: {result}')

if __name__ == '__main__':
    main()
```

❶ This function simply prints whatever arguments it gets, preceded by a timestamp in the format [HH:MM:SS].

❷ loiter does nothing except display a message when it starts, sleep for n seconds, then display a message when it ends; tabs are used to indent the messages according to the value of n.

❸ loiter returns n * 10 so we can see how to collect results.

❹ Create a ThreadPoolExecutor with three threads.

❺ Submit five tasks to the executor. Since there are only three threads, only three of those tasks will start immediately: the calls loiter(0), loiter(1), and loiter(2); this is a nonblocking call.

❻ Immediately display the results of invoking executor.map: it's a generator, as the output in Example 20-9 shows.

❼ The enumerate call in the for loop will implicitly invoke next(results), which in turn will invoke _f.result() on the (internal) _f future representing the first call, loiter(0). The result method will block until the future is done, therefore each iteration in this loop will have to wait for the next result to be ready.

I encourage you to run Example 20-8 and see the display being updated incrementally. While you're at it, play with the max_workers argument for the ThreadPoolExecutor and with the range function that produces the arguments for the executor.map call—or replace it with lists of handpicked values to create different delays.

Example 20-9 shows a sample run of Example 20-8.

Example 20-9. Sample run of demo_executor_map.py from Example 20-8

```
$ python3 demo_executor_map.py
[15:56:50] Script starting.  ❶
[15:56:50] loiter(0): doing nothing for 0s...  ❷
[15:56:50] loiter(0): done.
[15:56:50]     loiter(1): doing nothing for 1s...  ❸
[15:56:50]                 loiter(2): doing nothing for 2s...
[15:56:50] results: <generator object result_iterator at 0x106517168>  ❹
[15:56:50]                     loiter(3): doing nothing for 3s...  ❺
[15:56:50] Waiting for individual results:
[15:56:50] result 0: 0  ❻
[15:56:51]     loiter(1): done.  ❼
[15:56:51]                         loiter(4): doing nothing for 4s...
[15:56:51] result 1: 10  ❽
[15:56:52]             loiter(2): done.  ❾
[15:56:52] result 2: 20
[15:56:53]                 loiter(3): done.
[15:56:53] result 3: 30
[15:56:55]                     loiter(4): done.  ❿
[15:56:55] result 4: 40
```

❶ This run started at 15:56:50.

❷ The first thread executes loiter(0), so it will sleep for 0s and return even before the second thread has a chance to start, but YMMV.[6]

❸ loiter(1) and loiter(2) start immediately (because the thread pool has three workers, it can run three functions concurrently).

❹ This shows that the results returned by executor.map is a generator; nothing so far would block, regardless of the number of tasks and the max_workers setting.

---

6 Your mileage may vary: with threads, you never know the exact sequencing of events that should happen nearly at the same time; it's possible that, in another machine, you see loiter(1) starting before loiter(0) finishes, particularly because sleep always releases the GIL, so Python may switch to another thread even if you sleep for 0s.

❺ Because loiter(0) is done, the first worker is now available to start the fourth thread for loiter(3).

❻ This is where execution may block, depending on the parameters given to the loiter calls: the __next__ method of the results generator must wait until the first future is complete. In this case, it won't block because the call to loiter(0) finished before this loop started. Note that everything up to this point happened within the same second: 15:56:50.

❼ loiter(1) is done one second later, at 15:56:51. The thread is freed to start loiter(4).

❽ The result of loiter(1) is shown: 10. Now the for loop will block waiting for the result of loiter(2).

❾ The pattern repeats: loiter(2) is done, its result is shown; same with loiter(3).

❿ There is a 2s delay until loiter(4) is done, because it started at 15:56:51 and did nothing for 4s.

The Executor.map function is easy to use, but often it's preferable to get the results as they are ready, regardless of the order they were submitted. To do that, we need a combination of the Executor.submit method and the futures.as_completed function, as we saw in Example 20-4. We'll come back to this technique in "Using futures.as_completed" on page 773.

> The combination of executor.submit and futures.as_completed is more flexible than executor.map because you can submit different callables and arguments, while executor.map is designed to run the same callable on the different arguments. In addition, the set of futures you pass to futures.as_completed may come from more than one executor—perhaps some were created by a ThreadPoolExecutor instance, while others are from a ProcessPoolExecutor.

In the next section, we will resume the flag download examples with new requirements that will force us to iterate over the results of futures.as_completed instead of using executor.map.

Downloads with Progress Display and Error Handling

As mentioned, the scripts in "Concurrent Web Downloads" on page 748 have no error handling to make them easier to read and to contrast the structure of the three approaches: sequential, threaded, and asynchronous.

In order to test the handling of a variety of error conditions, I created the flags2 examples:

flags2_common.py

This module contains common functions and settings used by all flags2 examples, including a main function, which takes care of command-line parsing, timing, and reporting results. That is really support code, not directly relevant to the subject of this chapter, so I will not list the source code here, but you can read it in the *fluentpython/example-code-2e* repository: *20-executors/getflags/flags2_common.py* (*https://fpy.li/20-10*).

flags2_sequential.py

A sequential HTTP client with proper error handling and progress bar display. Its download_one function is also used by flags2_threadpool.py.

flags2_threadpool.py

Concurrent HTTP client based on futures.ThreadPoolExecutor to demonstrate error handling and integration of the progress bar.

flags2_asyncio.py

Same functionality as the previous example, but implemented with asyncio and httpx. This will be covered in "Enhancing the asyncio Downloader" on page 792, in Chapter 21.

Be Careful when Testing Concurrent Clients

When testing concurrent HTTP clients on public web servers, you may generate many requests per second, and that's how denial-of-service (DoS) attacks are made. Carefully throttle your clients when hitting public servers. For testing, set up a local HTTP server. See "Setting Up Test Servers" on page 769 for instructions.

The most visible feature of the flags2 examples is that they have an animated, text-mode progress bar implemented with the *tqdm* package (*https://fpy.li/20-11*). I posted a 108s video on YouTube (*https://fpy.li/20-12*) to show the progress bar and contrast the speed of the three flags2 scripts. In the video, I start with the sequential download, but I interrupt it after 32s because it was going to take more than 5 minutes to hit on 676 URLs and get 194 flags. I then run the threaded and asyncio scripts three times each, and every time they complete the job in 6s or less (i.e., more than 60 times faster). Figure 20-1 shows two screenshots: during and after running *flags2_threadpool.py*.

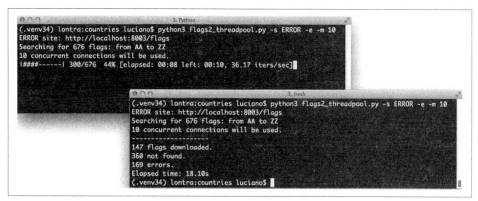

Figure 20-1. Top-left: flags2_threadpool.py running with live progress bar generated by tqdm; bottom-right: same terminal window after the script is finished.

The simplest *tqdm* example appears in an animated *.gif* in the project's *README.md* (*https://fpy.li/20-13*). If you type the following code in the Python console after installing the *tqdm* package, you'll see an animated progress bar where the comment is:

```
>>> import time
>>> from tqdm import tqdm
>>> for i in tqdm(range(1000)):
...     time.sleep(.01)
...
>>> # -> progress bar will appear here <-
```

Besides the neat effect, the tqdm function is also interesting conceptually: it consumes any iterable and produces an iterator which, while it's consumed, displays the progress bar and estimates the remaining time to complete all iterations. To compute that estimate, tqdm needs to get an iterable that has a len, or additionally receive the total= argument with the expected number of items. Integrating tqdm with our flags2 examples provides an opportunity to look deeper into how the concurrent scripts actually work, by forcing us to use the futures.as_completed (*https://fpy.li/ 20-7*) and the asyncio.as_completed (*https://fpy.li/20-15*) functions so that tqdm can display progress as each future is completed.

The other feature of the flags2 example is a command-line interface. All three scripts accept the same options, and you can see them by running any of the scripts with the -h option. Example 20-10 shows the help text.

Example 20-10. Help screen for the scripts in the flags2 series

```
$ python3 flags2_threadpool.py -h
usage: flags2_threadpool.py [-h] [-a] [-e] [-l N] [-m CONCURRENT] [-s LABEL]
                            [-v]
                            [CC [CC ...]]

Download flags for country codes. Default: top 20 countries by population.

positional arguments:
  CC                    country code or 1st letter (eg. B for BA...BZ)

optional arguments:
  -h, --help            show this help message and exit
  -a, --all             get all available flags (AD to ZW)
  -e, --every           get flags for every possible code (AA...ZZ)
  -l N, --limit N       limit to N first codes
  -m CONCURRENT, --max_req CONCURRENT
                        maximum concurrent requests (default=30)
  -s LABEL, --server LABEL
                        Server to hit; one of DELAY, ERROR, LOCAL, REMOTE
                        (default=LOCAL)
  -v, --verbose         output detailed progress info
```

All arguments are optional. But the `-s/--server` is essential for testing: it lets you choose which HTTP server and port will be used in the test. Pass one of these case-insensitive labels to determine where the script will look for the flags:

LOCAL

Use `http://localhost:8000/flags`; this is the default. You should configure a local HTTP server to answer at port 8000. See the following note for instructions.

REMOTE

Use `http://fluentpython.com/data/flags`; that is a public website owned by me, hosted on a shared server. Please do not pound it with too many concurrent requests. The *fluentpython.com* domain is handled by the Cloudflare (*https://fpy.li/20-16*) CDN (Content Delivery Network) so you may notice that the first downloads are slower, but they get faster when the CDN cache warms up.

DELAY

Use `http://localhost:8001/flags`; a server delaying HTTP responses should be listening to port 8001. I wrote *slow_server.py* to make it easier to experiment. You'll find it in the *20-futures/getflags/* directory of the *Fluent Python* code repository (*https://fpy.li/code*). See the following note for instructions.

ERROR

Use `http://localhost:8002/flags`; a server returning some HTTP errors should be listening on port 8002. Instructions are next.

Setting Up Test Servers

If you don't have a local HTTP server for testing, I wrote setup instructions using only Python ≥ 3.9 (no external libraries) in *20-executors/getflags/README.adoc* (*https://fpy.li/20-17*) in the *fluent-python/example-code-2e* (*https://fpy.li/code*) repository. In short, *README.adoc* describes how to use:

```
python3 -m http.server
```
> The LOCAL server on port 8000

```
python3 slow_server.py
```
> The DELAY server on port 8001, which adds a random delay of .5s to 5s before each response

```
python3 slow_server.py 8002 --error-rate .25
```
> The ERROR server on port 8002, which in addition to the random delay, has a 25% chance of returning a "418 I'm a teapot" (*https://fpy.li/20-18*) error response

By default, each *flags2*.py* script will fetch the flags of the 20 most populous countries from the LOCAL server (`http://localhost:8000/flags`) using a default number of concurrent connections, which varies from script to script. Example 20-11 shows a sample run of the *flags2_sequential.py* script using all defaults. To run it, you need a local server, as explained in "Be Careful when Testing Concurrent Clients" on page 766.

Example 20-11. Running flags2_sequential.py with all defaults: LOCAL site, top 20 flags, 1 concurrent connection

```
$ python3 flags2_sequential.py
LOCAL site: http://localhost:8000/flags
Searching for 20 flags: from BD to VN
1 concurrent connection will be used.
--------------------
20 flags downloaded.
Elapsed time: 0.10s
```

You can select which flags will be downloaded in several ways. Example 20-12 shows how to download all flags with country codes starting with the letters A, B, or C.

Example 20-12. Run flags2_threadpool.py to fetch all flags with country codes prefixes A, B, or C from the DELAY server

```
$ python3 flags2_threadpool.py -s DELAY a b c
DELAY site: http://localhost:8001/flags
Searching for 78 flags: from AA to CZ
```

```
30 concurrent connections will be used.
--------------------
43 flags downloaded.
35 not found.
Elapsed time: 1.72s
```

Regardless of how the country codes are selected, the number of flags to fetch can be limited with the -l/--limit option. Example 20-13 demonstrates how to run exactly 100 requests, combining the -a option to get all flags with -l 100.

Example 20-13. Run flags2_asyncio.py to get 100 flags (-al 100) from the ERROR server, using 100 concurrent requests (-m 100)

```
$ python3 flags2_asyncio.py -s ERROR -al 100 -m 100
ERROR site: http://localhost:8002/flags
Searching for 100 flags: from AD to LK
100 concurrent connections will be used.
--------------------
73 flags downloaded.
27 errors.
Elapsed time: 0.64s
```

That's the user interface of the flags2 examples. Let's see how they are implemented.

Error Handling in the flags2 Examples

The common strategy in all three examples to deal with HTTP errors is that 404 errors (not found) are handled by the function in charge of downloading a single file (download_one). Any other exception propagates to be handled by the down load_many function or the supervisor coroutine—in the asyncio example.

Once more, we'll start by studying the sequential code, which is easier to follow—and mostly reused by the thread pool script. Example 20-14 shows the functions that perform the actual downloads in the *flags2_sequential.py* and *flags2_threadpool.py* scripts.

Example 20-14. flags2_sequential.py: basic functions in charge of downloading; both are reused in flags2_threadpool.py

```
from collections import Counter
from http import HTTPStatus

import httpx
import tqdm  # type: ignore  ❶

from flags2_common import main, save_flag, DownloadStatus  ❷

DEFAULT_CONCUR_REQ = 1
```

```
MAX_CONCUR_REQ = 1

def get_flag(base_url: str, cc: str) -> bytes:
    url = f'{base_url}/{cc}/{cc}.gif'.lower()
    resp = httpx.get(url, timeout=3.1, follow_redirects=True)
    resp.raise_for_status()  ❸
    return resp.content

def download_one(cc: str, base_url: str, verbose: bool = False) -> DownloadStatus:
    try:
        image = get_flag(base_url, cc)
    except httpx.HTTPStatusError as exc:  ❹
        res = exc.response
        if res.status_code == HTTPStatus.NOT_FOUND:
            status = DownloadStatus.NOT_FOUND  ❺
            msg = f'not found: {res.url}'
        else:
            raise  ❻
    else:
        save_flag(image, f'{cc}.gif')
        status = DownloadStatus.OK
        msg = 'OK'

    if verbose:  ❼
        print(cc, msg)

    return status
```

❶ Import the tqdm progress-bar display library, and tell Mypy to skip checking it.[7]

❷ Import a couple of functions and an Enum from the flags2_common module.

❸ Raises HTTPStetusError if the HTTP status code is not in range(200, 300).

❹ download_one catches HTTPStatusError to handle HTTP code 404 specifically...

❺ ...by setting its local status to DownloadStatus.NOT_FOUND; DownloadStatus is an Enum imported from *flags2_common.py*.

❻ Any other HTTPStatusError exception is re-raised to propagate to the caller.

❼ If the -v/--verbose command-line option is set, the country code and status message are displayed; this is how you'll see progress in verbose mode.

---

7 As of September 2021, there are no type hints in the current release of tdqm. That's OK. The world will not end because of that. Thank Guido for optional typing!

Example 20-15 lists the sequential version of the download_many function. This code is straightforward, but it's worth studying to contrast with the concurrent versions coming up. Focus on how it reports progress, handles errors, and tallies downloads.

Example 20-15. flags2_sequential.py: the sequential implementation of download_many

```
def download_many(cc_list: list[str],
                  base_url: str,
                  verbose: bool,
                  _unused_concur_req: int) -> Counter[DownloadStatus]:
    counter: Counter[DownloadStatus] = Counter()  ❶
    cc_iter = sorted(cc_list)  ❷
    if not verbose:
        cc_iter = tqdm.tqdm(cc_iter)  ❸
    for cc in cc_iter:
        try:
            status = download_one(cc, base_url, verbose)  ❹
        except httpx.HTTPStatusError as exc:  ❺
            error_msg = 'HTTP error {resp.status_code} - {resp.reason_phrase}'
            error_msg = error_msg.format(resp=exc.response)
        except httpx.RequestError as exc:  ❻
            error_msg = f'{exc} {type(exc)}'.strip()
        except KeyboardInterrupt:  ❼
            break
        else:  ❽
            error_msg = ''

        if error_msg:
            status = DownloadStatus.ERROR  ❾
        counter[status] += 1           ❿
        if verbose and error_msg:      ⓫
            print(f'{cc} error: {error_msg}')

    return counter  ⓬
```

❶ This Counter will tally the different download outcomes: DownloadStatus.OK, DownloadStatus.NOT_FOUND, or DownloadStatus.ERROR.

❷ cc_iter holds the list of the country codes received as arguments, ordered alphabetically.

❸ If not running in verbose mode, cc_iter is passed to tqdm, which returns an iterator yielding the items in cc_iter while also animating the progress bar.

❹ Make successive calls to download_one.

❺ HTTP status code exceptions raised by `get_flag` and not handled by `download_one` are handled here.

❻ Other network-related exceptions are handled here. Any other exception will abort the script, because the `flags2_common.main` function that calls `download_many` has no try/except.

❼ Exit the loop if the user hits Ctrl-C.

❽ If no exception escaped `download_one`, clear the error message.

❾ If there was an error, set the local `status` accordingly.

❿ Increment the counter for that `status`.

⓫ In verbose mode, display the error message for the current country code, if any.

⓬ Return `counter` so that `main` can display the numbers in the final report.

We'll now study the refactored thread pool example, *flags2_threadpool.py*.

Using futures.as_completed

In order to integrate the *tqdm* progress bar and handle errors on each request, the *flags2_threadpool.py* script uses `futures.ThreadPoolExecutor` with the `futures.as_completed` function we've already seen. Example 20-16 is the full listing of *flags2_threadpool.py*. Only the `download_many` function is implemented; the other functions are reused from *flags2_common.py* and *flags2_sequential.py*.

Example 20-16. flags2_threadpool.py: full listing

```
from collections import Counter
from concurrent.futures import ThreadPoolExecutor, as_completed

import httpx
import tqdm  # type: ignore

from flags2_common import main, DownloadStatus
from flags2_sequential import download_one  ❶

DEFAULT_CONCUR_REQ = 30  ❷
MAX_CONCUR_REQ = 1000  ❸


def download_many(cc_list: list[str],
                  base_url: str,
```

```
                verbose: bool,
                concur_req: int) -> Counter[DownloadStatus]:
    counter: Counter[DownloadStatus] = Counter()
    with ThreadPoolExecutor(max_workers=concur_req) as executor:    ❹
        to_do_map = {}  ❺
        for cc in sorted(cc_list):  ❻
            future = executor.submit(download_one, cc,
                                     base_url, verbose)  ❼
            to_do_map[future] = cc  ❽
        done_iter = as_completed(to_do_map)  ❾
        if not verbose:
            done_iter = tqdm.tqdm(done_iter, total=len(cc_list))  ❿
        for future in done_iter:  ⓫
            try:
                status = future.result()  ⓬
            except httpx.HTTPStatusError as exc:  ⓭
                error_msg = 'HTTP error {resp.status_code} - {resp.reason_phrase}'
                error_msg = error_msg.format(resp=exc.response)
            except httpx.RequestError as exc:
                error_msg = f'{exc} {type(exc)}'.strip()
            except KeyboardInterrupt:
                break
            else:
                error_msg = ''

            if error_msg:
                status = DownloadStatus.ERROR
            counter[status] += 1
            if verbose and error_msg:
                cc = to_do_map[future]  ⓮
                print(f'{cc} error: {error_msg}')

    return counter


if __name__ == '__main__':
    main(download_many, DEFAULT_CONCUR_REQ, MAX_CONCUR_REQ)
```

❶ Reuse download_one from flags2_sequential (Example 20-14).

❷ If the -m/--max_req command-line option is not given, this will be the maximum number of concurrent requests, implemented as the size of the thread pool; the actual number may be smaller if the number of flags to download is smaller.

❸ MAX_CONCUR_REQ caps the maximum number of concurrent requests regardless of the number of flags to download or the -m/--max_req command-line option. It's a safety precaution to avoid launching too many threads with their significant memory overhead.

❹ Create the executor with max_workers set to concur_req, computed by the main function as the smaller of: MAX_CONCUR_REQ, the length of cc_list, or the value of the -m/--max_req command-line option. This avoids creating more threads than necessary.

❺ This dict will map each Future instance—representing one download—with the respective country code for error reporting.

❻ Iterate over the list of country codes in alphabetical order. The order of the results will depend on the timing of the HTTP responses more than anything, but if the size of the thread pool (given by concur_req) is much smaller than len(cc_list), you may notice the downloads batched alphabetically.

❼ Each call to executor.submit schedules the execution of one callable and returns a Future instance. The first argument is the callable, the rest are the arguments it will receive.

❽ Store the future and the country code in the dict.

❾ futures.as_completed returns an iterator that yields futures as each task is done.

❿ If not in verbose mode, wrap the result of as_completed with the tqdm function to display the progress bar; because done_iter has no len, we must tell tqdm what is the expected number of items as the total= argument, so tqdm can estimate the work remaining.

⓫ Iterate over the futures as they are completed.

⓬ Calling the result method on a future either returns the value returned by the callable, or raises whatever exception was caught when the callable was executed. This method may block waiting for a resolution, but not in this example because as_completed only returns futures that are done.

⓭ Handle the potential exceptions; the rest of this function is identical to the sequential download_many in Example 20-15), except for the next callout.

⓮ To provide context for the error message, retrieve the country code from the to_do_map using the current future as key. This was not necessary in the sequential version because we were iterating over the list of country codes, so we knew the current cc; here we are iterating over the futures.

Example 20-16 uses an idiom that is very useful with futures.as_completed: building a dict to map each future to other data that may be useful when the future is completed. Here the to_do_map maps each future to the country code assigned to it. This makes it easy to do follow-up processing with the result of the futures, despite the fact that they are produced out of order.

Python threads are well suited for I/O-intensive applications, and the concur rent.futures package makes it relatively simple to use for certain use cases. With ProcessPoolExecutor, you can also solve CPU-intensive problems on multiple cores —if the computations are "embarrassingly parallel" (*https://fpy.li/20-19*). This concludes our basic introduction to concurrent.futures.

Chapter Summary

We started the chapter by comparing two concurrent HTTP clients with a sequential one, demonstrating that the concurrent solutions show significant performance gains over the sequential script.

After studying the first example based on concurrent.futures, we took a closer look at future objects, either instances of concurrent.futures.Future or asyncio .Future, emphasizing what these classes have in common (their differences will be emphasized in Chapter 21). We saw how to create futures by calling Executor.sub mit, and iterate over completed futures with concurrent.futures.as_completed.

We then discussed the use of multiple processes with the concurrent.futures.Proc essPoolExecutor class, to go around the GIL and use multiple CPU cores to simplify the multicore prime checker we first saw in Chapter 19.

In the following section, we saw how the concurrent.futures.ThreadPoolExecutor works with a didactic example, launching tasks that did nothing for a few seconds, except for displaying their status with a timestamp.

Next we went back to the flag downloading examples. Enhancing them with a progress bar and proper error handling prompted further exploration of the future.as_completed generator function, showing a common pattern: storing futures in a dict to link further information to them when submitting, so that we can use that information when the future comes out of the as_completed iterator.

Further Reading

The concurrent.futures package was contributed by Brian Quinlan, who presented it in a great talk titled "The Future Is Soon!" (*https://fpy.li/20-20*) at PyCon Australia 2010. Quinlan's talk has no slides; he shows what the library does by typing code directly in the Python console. As a motivating example, the presentation features a short video with XKCD cartoonist/programmer Randall Munroe making an unintended DoS attack on Google Maps to build a colored map of driving times around his city. The formal introduction to the library is PEP 3148 - `futures` - execute computations asynchronously (*https://fpy.li/pep3148*). In the PEP, Quinlan wrote that the `con current.futures` library was "heavily influenced by the Java `java.util.concurrent` package."

For additional resources covering `concurrent.futures`, please see Chapter 19. All the references that cover Python's `threading` and `multiprocessing` in "Concurrency with Threads and Processes" on page 738 also cover `concurrent.futures`.

Soapbox

Thread Avoidance

> Concurrency: one of the most difficult topics in computer science (usually best avoided).
>
> —David Beazley, Python instructor and mad scientist[8]

I agree with the apparently contradictory quotes by David Beazley and Michele Simionato at the start of this chapter.

I attended an undergraduate course about concurrency. All we did was POSIX threads (*https://fpy.li/20-22*) programming. What I learned: I don't want to manage threads and locks myself, for the same reason that I don't want to manage memory allocation and deallocation. Those jobs are best carried out by the systems programmers who have the know-how, the inclination, and the time to get them right—hopefully. I am paid to develop applications, not operating systems. I don't need all the fine-grained control of threads, locks, `malloc`, and `free`—see "C dynamic memory allocation" (*https://fpy.li/20-23*).

That's why I think the `concurrent.futures` package is interesting: it treats threads, processes, and queues as infrastructure at your service, not something you have to deal with directly. Of course, it's designed with simple jobs in mind, the so-called embarrassingly parallel problems. But that's a large slice of the concurrency problems we face when writing applications—as opposed to operating systems or database servers, as Simionato points out in that quote.

For "nonembarrassing" concurrency problems, threads and locks are not the answer either. Threads will never disappear at the OS level, but every programming language

---

8 Slide #9 from "A Curious Course on Coroutines and Concurrency" (*https://fpy.li/20-21*) tutorial presented at PyCon 2009.

I've found exciting in the last several years provides higher-level, concurrency abstractions that are easier to use correctly, as the excellent *Seven Concurrency Models in Seven Weeks (https://fpy.li/20-24)* book by Paul Butcher demonstrates. Go, Elixir, and Clojure are among them. Erlang—the implementation language of Elixir—is a prime example of a language designed from the ground up with concurrency in mind. Erlang doesn't excite me for a simple reason: I find its syntax ugly. Python spoiled me that way.

José Valim, previously a Ruby on Rails core contributor, designed Elixir with a pleasant, modern syntax. Like Lisp and Clojure, Elixir implements syntactic macros. That's a double-edged sword. Syntactic macros enable powerful DSLs, but the proliferation of sublanguages can lead to incompatible codebases and community fragmentation. Lisp drowned in a flood of macros, with each Lisp shop using its own arcane dialect. Standardizing around Common Lisp resulted in a bloated language. I hope José Valim can inspire the Elixir community to avoid a similar outcome. So far, it's looking good. The Ecto (*https://fpy.li/20-25*) database wrapper and query generator is a joy to use: a great example of using macros to create a flexible yet user-friendly DSL—Domain-Specific Language—for interacting with relational and nonrelational databases.

Like Elixir, Go is a modern language with fresh ideas. But, in some regards, it's a conservative language, compared to Elixir. Go doesn't have macros, and its syntax is simpler than Python's. Go doesn't support inheritance or operator overloading, and it offers fewer opportunities for metaprogramming than Python. These limitations are considered features. They lead to more predictable behavior and performance. That's a big plus in the highly concurrent, mission-critical settings where Go aims to replace C++, Java, and Python.

While Elixir and Go are direct competitors in the high-concurrency space, their design philosophies appeal to different crowds. Both are likely to thrive. But in the history of programming languages, the conservative ones tend to attract more coders.

## CHAPTER 21

Asynchronous Programming

> The problem with normal approaches to asynchronous programming is that they're all-or-nothing propositions. You rewrite all your code so none of it blocks or you're just wasting your time.
>
> Alvaro Videla and Jason J. W. Williams, *RabbitMQ in Action*[1]

This chapter addresses three major topics that are closely related:

- Python's `async def`, `await`, `async with`, and `async for` constructs
- Objects supporting those constructs: native coroutines and asynchronous variants of context managers, iterables, generators, and comprehensions
- *asyncio* and other asynchronous libraries

This chapter builds on the ideas of iterables and generators (Chapter 17, in particular "Classic Coroutines" on page 645), context managers (Chapter 18), and general concepts of concurrent programming (Chapter 19).

We'll study concurrent HTTP clients similar to the ones we saw in Chapter 20, rewritten with native coroutines and asynchronous context managers, using the same *HTTPX* library as before, but now through its asynchronous API. We'll also see how to avoid blocking the event loop by delegating slow operations to a thread or process executor.

After the HTTP client examples, we'll see two simple asynchronous server-side applications, one of them using the increasingly popular *FastAPI* framework. Then we'll cover other language constructs enabled by the `async`/`await` keywords:

---

1 Videla & Williams, *RabbitMQ in Action* (Manning), Chapter 4, "Solving Problems with Rabbit: coding and patterns," p. 61.

asynchronous generator functions, asynchronous comprehensions, and asynchronous generator expressions. To emphasize the fact that those language features are not tied to *asyncio*, we'll see one example rewritten to use *Curio*—the elegant and innovative asynchronous framework invented by David Beazley.

To wrap up the chapter, I wrote a brief section on the advantages and pitfalls of asynchronous programming.

That's a lot of ground to cover. We only have space for basic examples, but they will illustrate the most important features of each idea.

 The *asyncio* documentation (*https://fpy.li/21-1*) is much better after Yury Selivanov[2] reorganized it, separating the few functions useful to application developers from the low-level API for creators of packages like web frameworks and database drivers.

For book-length coverage of *asyncio*, I recommend *Using Asyncio in Python* by Caleb Hattingh (O'Reilly). Full disclosure: Caleb is one of the tech reviewers of this book.

What's New in This Chapter

When I wrote the first edition of *Fluent Python*, the *asyncio* library was provisional and the `async`/`await` keywords did not exist. Therefore, I had to update all examples in this chapter. I also created new examples: domain probing scripts, a *FastAPI* web service, and experiments with Python's new asynchronous console mode.

New sections cover language features that did not exist at the time, such as native coroutines, `async with`, `async for`, and the objects that support those constructs.

The ideas in "How Async Works and How It Doesn't" on page 829 reflect hard-earned lessons that I consider essential reading for anyone using asynchronous programming. They may save you a lot of trouble—whether you're using Python or Node.js.

Finally, I removed several paragraphs about `asyncio.Futures`, which is now considered part of the low-level *asyncio* APIs.

---

2 Selivanov implemented `async`/`await` in Python, and wrote the related PEPs 492 (*https://fpy.li/pep492*), 525 (*https://fpy.li/pep525*), and 530 (*https://fpy.li/pep530*).

A Few Definitions

At the start of "Classic Coroutines" on page 645, we saw that Python 3.5 and later offer three kinds of coroutines:

Native coroutine
> A coroutine function defined with `async def`. You can delegate from a native coroutine to another native coroutine using the `await` keyword, similar to how classic coroutines use `yield from`. The `async def` statement always defines a native coroutine, even if the `await` keyword is not used in its body. The `await` keyword cannot be used outside of a native coroutine.[3]

Classic coroutine
> A generator function that consumes data sent to it via `my_coro.send(data)` calls, and reads that data by using `yield` in an expression. Classic coroutines can delegate to other classic coroutines using `yield from`. Classic coroutines cannot be driven by `await`, and are no longer supported by *asyncio*.

Generator-based coroutine
> A generator function decorated with `@types.coroutine`—introduced in Python 3.5. That decorator makes the generator compatible with the new `await` keyword.

In this chapter, we focus on native coroutines as well as *asynchronous generators*:

Asynchronous generator
> A generator function defined with `async def` and using `yield` in its body. It returns an asynchronous generator object that provides `__anext__`, a coroutine method to retrieve the next item.

> **@asyncio.coroutine has No Future[4]**
>
> The `@asyncio.coroutine` decorator for classic coroutines and generator-based coroutines was deprecated in Python 3.8 and is scheduled for removal in Python 3.11, according to Issue 43216 (*https://fpy.li/21-2*). In contrast, `@types.coroutine` should remain, per Issue 36921 (*https://fpy.li/21-3*). It is no longer supported by *asyncio*, but is used in low-level code in the *Curio* and *Trio* asynchronous frameworks.

---

3 There is one exception to this rule: if you run Python with the `-m asyncio` option, you can use `await` directly at the >>> prompt to drive a native coroutine. This is explained in "Experimenting with Python's async console" on page 816.

4 Sorry, I could not resist it.

An asyncio Example: Probing Domains

Imagine you are about to start a new blog on Python, and you plan to register a domain using a Python keyword and the *.DEV* suffix—for example: *AWAIT.DEV*. Example 21-1 is a script using *asyncio* to check several domains concurrently. This is the output it produces:

```
$ python3 blogdom.py
  with.dev
+ elif.dev
+ def.dev
  from.dev
  else.dev
  or.dev
  if.dev
  del.dev
+ as.dev
  none.dev
  pass.dev
  true.dev
+ in.dev
+ for.dev
+ is.dev
+ and.dev
+ try.dev
+ not.dev
```

Note that the domains appear unordered. If you run the script, you'll see them displayed one after the other, with varying delays. The + sign indicates your machine was able to resolve the domain via DNS. Otherwise, the domain did not resolve and may be available.[5]

In *blogdom.py*, the DNS probing is done via native coroutine objects. Because the asynchronous operations are interleaved, the time needed to check the 18 domains is much less than checking them sequentially. In fact, the total time is practically the same as the time for the single slowest DNS response, instead of the sum of the times of all responses.

Example 21-1 shows the code for *blogdom.py*.

Example 21-1. blogdom.py: search for domains for a Python blog

```
#!/usr/bin/env python3
import asyncio
import socket
```

---

5 true.dev is available for USD 360/year as I write this. I see that for.dev is registered, but has no DNS configured.

```
from keyword import kwlist

MAX_KEYWORD_LEN = 4  ❶


async def probe(domain: str) -> tuple[str, bool]:  ❷
    loop = asyncio.get_running_loop()  ❸
    try:
        await loop.getaddrinfo(domain, None)  ❹
    except socket.gaierror:
        return (domain, False)
    return (domain, True)


async def main() -> None:  ❺
    names = (kw for kw in kwlist if len(kw) <= MAX_KEYWORD_LEN)  ❻
    domains = (f'{name}.dev'.lower() for name in names)  ❼
    coros = [probe(domain) for domain in domains]  ❽
    for coro in asyncio.as_completed(coros):  ❾
        domain, found = await coro  ❿
        mark = '+' if found else ' '
        print(f'{mark} {domain}')


if __name__ == '__main__':
    asyncio.run(main())  ⓫
```

❶ Set maximum length of keyword for domains, because shorter is better.

❷ probe returns a tuple with the domain name and a boolean; True means the domain resolved. Returning the domain name will make it easier to display the results.

❸ Get a reference to the asyncio event loop, so we can use it next.

❹ The loop.getaddrinfo(…) (https://fpy.li/21-4) coroutine-method returns a five-part tuple of parameters (https://fpy.li/21-5) to connect to the given address using a socket. In this example, we don't need the result. If we got it, the domain resolves; otherwise, it doesn't.

❺ main must be a coroutine, so that we can use await in it.

❻ Generator to yield Python keywords with length up to MAX_KEYWORD_LEN.

❼ Generator to yield domain names with the .dev suffix.

❽ Build a list of coroutine objects by invoking the `probe` coroutine with each `domain` argument.

❾ `asyncio.as_completed` is a generator that yields coroutines that return the results of the coroutines passed to it in the order they are completed—not the order they were submitted. It's similar to `futures.as_completed`, which we saw in Chapter 20, Example 20-4.

❿ At this point, we know the coroutine is done because that's how `as_completed` works. Therefore, the `await` expression will not block but we need it to get the result from `coro`. If `coro` raised an unhandled exception, it would be re-raised here.

⓫ `asyncio.run` starts the event loop and returns only when the event loop exits. This is a common pattern for scripts that use `asyncio`: implement `main` as a coroutine, and drive it with `asyncio.run` inside the `if __name__ == '__main__':` block.

> The `asyncio.get_running_loop` function was added in Python 3.7 for use inside coroutines, as shown in `probe`. If there's no running loop, `asyncio.get_running_loop` raises `RuntimeError`. Its implementation is simpler and faster than `asyncio.get_event_loop`, which may start an event loop if necessary. Since Python 3.10, `asyncio.get_event_loop` is deprecated (*https://fpy.li/21-6*) and will eventually become an alias to `asyncio.get_running_loop`.

Guido's Trick to Read Asynchronous Code

There are a lot of new concepts to grasp in *asyncio*, but the overall logic of Example 21-1 is easy to follow if you employ a trick suggested by Guido van Rossum himself: squint and pretend the `async` and `await` keywords are not there. If you do that, you'll realize that coroutines read like plain old sequential functions.

For example, imagine that the body of this coroutine…

```python
async def probe(domain: str) -> tuple[str, bool]:
    loop = asyncio.get_running_loop()
    try:
        await loop.getaddrinfo(domain, None)
    except socket.gaierror:
        return (domain, False)
    return (domain, True)
```

…works like the following function, except that it magically never blocks:

```
def probe(domain: str) -> tuple[str, bool]:  # no async
    loop = asyncio.get_running_loop()
    try:
        loop.getaddrinfo(domain, None)  # no await
    except socket.gaierror:
        return (domain, False)
    return (domain, True)
```

Using the syntax await loop.getaddrinfo(...) avoids blocking because await suspends the current coroutine object. For example, during the execution of the probe('if.dev') coroutine, a new coroutine object is created by getad drinfo('if.dev', None). Awaiting it starts the low-level addrinfo query and yields control back to the event loop, not to the probe('if.dev') coroutine, which is suspended. The event loop can then drive other pending coroutine objects, such as probe('or.dev').

When the event loop gets a response for the getaddrinfo('if.dev', None) query, that specific coroutine object resumes and returns control back to the probe('if.dev')—which was suspended at await—and can now handle a possible exception and return the result tuple.

So far, we've only seen asyncio.as_completed and await applied to coroutines. But they handle any *awaitable* object. That concept is explained next.

New Concept: Awaitable

The for keyword works with *iterables*. The await keyword works with *awaitables*.

As an end user of *asyncio*, these are the awaitables you will see on a daily basis:

- A *native coroutine object*, which you get by calling a *native coroutine function*
- An asyncio.Task, which you usually get by passing a coroutine object to asyn cio.create_task()

However, end-user code does not always need to await on a Task. We use asyn cio.create_task(one_coro()) to schedule one_coro for concurrent execution, without waiting for its return. That's what we did with the spinner coroutine in *spinner_async.py* (Example 19-4). If you don't expect to cancel the task or wait for it, there is no need to keep the Task object returned from create_task. Creating the task is enough to schedule the coroutine to run.

In contrast, we use await other_coro() to run other_coro right now and wait for its completion because we need its result before we can proceed. In *spinner_async.py*, the supervisor coroutine did res = await slow() to execute slow and get its result.

When implementing asynchronous libraries or contributing to *asyncio* itself, you may also deal with these lower-level awaitables:

- An object with an __await__ method that returns an iterator; for example, an asyncio.Future instance (asyncio.Task is a subclass of asyncio.Future)
- Objects written in other languages using the Python/C API with a tp_as_async.am_await function, returning an iterator (similar to __await__ method)

Existing codebases may also have one additional kind of awaitable: *generator-based coroutine objects*—which are in the process of being deprecated.

 PEP 492 states (*https://fpy.li/21-7*) that the await expression "uses the yield from implementation with an extra step of validating its argument" and "await only accepts an awaitable." The PEP does not explain that implementation in detail, but refers to PEP 380 (*https://fpy.li/pep380*), which introduced yield from. I posted a detailed explanation in "Classic Coroutines" (*https://fpy.li/oldcoro*), section "The Meaning of yield from" (*https://fpy.li/21-8*), at *fluent-python.com*.

Now let's study the *asyncio* version of a script that downloads a fixed set of flag images.

Downloading with asyncio and HTTPX

The *flags_asyncio.py* script downloads a fixed set of 20 flags from *fluentpython.com*. We first mentioned it in "Concurrent Web Downloads" on page 748, but now we'll study it in detail, applying the concepts we just saw.

As of Python 3.10, *asyncio* only supports TCP and UDP directly, and there are no asynchronous HTTP client or server packages in the standard library. I am using HTTPX (*https://fpy.li/httpx*) in all the HTTP client examples.

We'll explore *flags_asyncio.py* from the bottom up—that is, looking first at the functions that set up the action in Example 21-2.

To make the code easier to read, *flags_asyncio.py* has no error handling. As we introduce async/await, it's useful to focus on the "happy path" initially, to understand how regular functions and coroutines are arranged in a program. Starting with "Enhancing the asyncio Downloader" on page 792, the examples include error handling and more features.

The *flags_.py* examples from this chapter and Chapter 20 share code and data, so I put them together in the *example-code-2e/20-executors/getflags* (*https://fpy.li/21-9*) directory.

Example 21-2. flags_asyncio.py: startup functions

```python
def download_many(cc_list: list[str]) -> int:      ❶
    return asyncio.run(supervisor(cc_list))        ❷

async def supervisor(cc_list: list[str]) -> int:
    async with AsyncClient() as client:            ❸
        to_do = [download_one(client, cc)
                    for cc in sorted(cc_list)]      ❹
        res = await asyncio.gather(*to_do)          ❺

    return len(res)                                 ❻

if __name__ == '__main__':
    main(download_many)
```

❶ This needs to be a plain function—not a coroutine—so it can be passed to and called by the main function from the *flags.py* module (Example 20-2).

❷ Execute the event loop driving the supervisor(cc_list) coroutine object until it returns. This will block while the event loop runs. The result of this line is whatever supervisor returns.

❸ Asynchronous HTTP client operations in httpx are methods of AsyncClient, which is also an asynchronous context manager: a context manager with asynchronous setup and teardown methods (more about this in "Asynchronous Context Managers" on page 790).

❹ Build a list of coroutine objects by calling the download_one coroutine once for each flag to be retrieved.

❺ Wait for the asyncio.gather coroutine, which accepts one or more awaitable arguments and waits for all of them to complete, returning a list of results for the given awaitables in the order they were submitted.

❻ supervisor returns the length of the list returned by `asyncio.gather`.

Now let's review the top of *flags_asyncio.py* (Example 21-3). I reorganized the coroutines so we can read them in the order they are started by the event loop.

Example 21-3. flags_asyncio.py: imports and download functions

```python
import asyncio

from httpx import AsyncClient  ❶

from flags import BASE_URL, save_flag, main  ❷

async def download_one(client: AsyncClient, cc: str):  ❸
    image = await get_flag(client, cc)
    save_flag(image, f'{cc}.gif')
    print(cc, end=' ', flush=True)
    return cc

async def get_flag(client: AsyncClient, cc: str) -> bytes:  ❹
    url = f'{BASE_URL}/{cc}/{cc}.gif'.lower()
    resp = await client.get(url, timeout=6.1,
                            follow_redirects=True)  ❺
    return resp.read()  ❻
```

❶ httpx must be installed—it's not in the standard library.

❷ Reuse code from *flags.py* (Example 20-2).

❸ download_one must be a native coroutine, so it can await on get_flag—which does the HTTP request. Then it displays the code of the downloaded flag, and saves the image.

❹ get_flag needs to receive the AsyncClient to make the request.

❺ The get method of an httpx.AsyncClient instance returns a ClientResponse object that is also an asynchronous context manager.

❻ Network I/O operations are implemented as coroutine methods, so they are driven asynchronously by the asyncio event loop.

For better performance, the `save_flag` call inside `get_flag` should be asynchronous, to avoid blocking the event loop. However, *asyncio* does not provide an asynchronous filesystem API at this time—as Node.js does.

"Using asyncio.as_completed and a Thread" on page 793 will show how to delegate `save_flag` to a thread.

Your code delegates to the `httpx` coroutines explicitly through `await` or implicitly through the special methods of the asynchronous context managers, such as `Async Client` and `ClientResponse`—as we'll see in "Asynchronous Context Managers" on page 790.

The Secret of Native Coroutines: Humble Generators

A key difference between the classic coroutine examples we saw in "Classic Coroutines" on page 645 and *flags_asyncio.py* is that there are no visible `.send()` calls or `yield` expressions in the latter. Your code sits between the *asyncio* library and the asynchronous libraries you are using, such as *HTTPX*. This is illustrated in Figure 21-1.

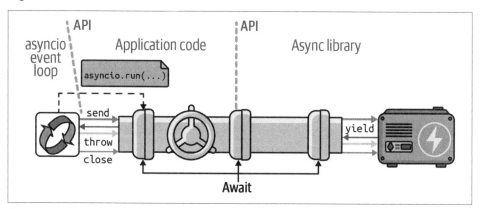

Figure 21-1. In an asynchronous program, a user's function starts the event loop, scheduling an initial coroutine with `asyncio.run`. Each user's coroutine drives the next with an `await` expression, forming a channel that enables communication between a library like HTTPX and the event loop.

Under the hood, the `asyncio` event loop makes the `.send` calls that drive your coroutines, and your coroutines `await` on other coroutines, including library coroutines. As mentioned, `await` borrows most of its implementation from `yield from`, which also makes `.send` calls to drive coroutines.

The await chain eventually reaches a low-level awaitable, which returns a generator that the event loop can drive in response to events such as timers or network I/O. The low-level awaitables and generators at the end of these await chains are implemented deep into the libraries, are not part of their APIs, and may be Python/C extensions.

Using functions like asyncio.gather and asyncio.create_task, you can start multiple concurrent await channels, enabling concurrent execution of multiple I/O operations driven by a single event loop, in a single thread.

The All-or-Nothing Problem

Note that in Example 21-3, I could not reuse the get_flag function from *flags.py* (Example 20-2). I had to rewrite it as a coroutine to use the asynchronous API of *HTTPX*. For peak performance with *asyncio*, we must replace every function that does I/O with an asynchronous version that is activated with await or asyncio.cre ate_task, so that control is given back to the event loop while the function waits for I/O. If you can't rewrite a blocking function as a coroutine, you should run it in a separate thread or process, as we'll see in "Delegating Tasks to Executors" on page 801.

That's why I chose the epigraph for this chapter, which includes this advice: "You rewrite all your code so none of it blocks or you're just wasting your time."

For the same reason, I could not reuse the download_one function from *flags_thread-pool.py* (Example 20-3) either. The code in Example 21-3 drives get_flag with await, so download_one must also be a coroutine. For each request, a download_one coroutine object is created in supervisor, and they are all driven by the asyncio.gather coroutine.

Now let's study the async with statement that appeared in supervisor (Example 21-2) and get_flag (Example 21-3).

Asynchronous Context Managers

In "Context Managers and with Blocks" on page 662, we saw how an object can be used to run code before and after the body of a with block, if its class provides the __enter__ and __exit__ methods.

Now, consider Example 21-4, from the *asyncpg* (*https://fpy.li/21-10*) *asyncio*-compatible PostgreSQL driver documentation on transactions (*https://fpy.li/21-11*).

Example 21-4. Sample code from the documentation of the asyncpg PostgreSQL driver

```
tr = connection.transaction()
await tr.start()
try:
```

```
    await connection.execute("INSERT INTO mytable VALUES (1, 2, 3)")
except:
    await tr.rollback()
    raise
else:
    await tr.commit()
```

A database transaction is a natural fit for the context manager protocol: the transaction has to be started, data is changed with `connection.execute`, and then a rollback or commit must happen, depending on the outcome of the changes.

In an asynchronous driver like *asyncpg*, the setup and wrap-up need to be coroutines so that other operations can happen concurrently. However, the implementation of the classic `with` statement doesn't support coroutines doing the work of `__enter__` or `__exit__`.

That's why PEP 492—Coroutines with async and await syntax (*https://fpy.li/pep492*) introduced the `async with` statement, which works with asynchronous context managers: objects implementing the `__aenter__` and `__aexit__` methods as coroutines.

With `async with`, Example 21-4 can be written like this other snippet from the *asyncpg* documentation (*https://fpy.li/21-11*):

```
async with connection.transaction():
    await connection.execute("INSERT INTO mytable VALUES (1, 2, 3)")
```

In the `asyncpg.Transaction` class (*https://fpy.li/21-13*), the `__aenter__` coroutine method does `await self.start()`, and the `__aexit__` coroutine awaits on private `__rollback` or `__commit` coroutine methods, depending on whether an exception occurred or not. Using coroutines to implement `Transaction` as an asynchronous context manager allows *asyncpg* to handle many transactions concurrently.

Caleb Hattingh on asyncpg

Another really great thing about *asyncpg* is that it also works around PostgreSQL's lack of high-concurrency support (it uses one server-side process per connection) by implementing a connection pool for internal connections to Postgres itself.

This means you don't need additional tools like *pgbouncer* as explained in the *asyncpg* documentation (*https://fpy.li/21-14*).[6]

---

6 This tip is quoted verbatim from a comment by tech reviewer Caleb Hattingh. Thanks, Caleb!

Back to *flags_asyncio.py*, the AsyncClient class of httpx is an asynchronous context manager, so it can use awaitables in its __aenter__ and __aexit__ special coroutine methods.

 "Asynchronous generators as context managers" on page 821 shows how to use Python's contextlib to create an asynchronous context manager without having to write a class. That explanation comes later in this chapter because of a prerequisite topic: "Asynchronous Generator Functions" on page 816.

We'll now enhance the *asyncio* flag download example with a progress bar, which will lead us to explore a bit more of the *asyncio* API.

Enhancing the asyncio Downloader

Recall from "Downloads with Progress Display and Error Handling" on page 766 that the flags2 set of examples share the same command-line interface, and they display a progress bar while the downloads are happening. They also include error handling.

 I encourage you to play with the flags2 examples to develop an intuition of how concurrent HTTP clients perform. Use the -h option to see the help screen in Example 20-10. Use the -a, -e, and -l command-line options to control the number of downloads, and the -m option to set the number of concurrent downloads. Run tests against the LOCAL, REMOTE, DELAY, and ERROR servers. Discover the optimum number of concurrent downloads to maximize throughput against each server. Tweak the options for the test servers, as described in "Setting Up Test Servers" on page 769.

For instance, Example 21-5 shows an attempt to get 100 flags (-al 100) from the ERROR server, using 100 concurrent requests (-m 100). The 48 errors in the result are either HTTP 418 or time-out errors—the expected (mis)behavior of the *slow_server.py*.

Example 21-5. Running flags2_asyncio.py

```
$ python3 flags2_asyncio.py -s ERROR -al 100 -m 100
ERROR site: http://localhost:8002/flags
Searching for 100 flags: from AD to LK
100 concurrent connections will be used.
100%|████████████████████████████| 100/100 [00:03<00:00, 30.48it/s]
--------------------
 52 flags downloaded.
 48 errors.
Elapsed time: 3.31s
```

Act Responsibly When Testing Concurrent Clients

Even if the overall download time is not much different between the threaded and *asyncio* HTTP clients, *asyncio* can send requests faster, so it's more likely that the server will suspect a DoS attack. To really exercise these concurrent clients at full throttle, please use local HTTP servers for testing, as explained in "Setting Up Test Servers" on page 769.

Now let's see how *flags2_asyncio.py* is implemented.

Using asyncio.as_completed and a Thread

In Example 21-3, we passed several coroutines to `asyncio.gather`, which returns a list with results of the coroutines in the order they were submitted. This means that `asyncio.gather` can only return when all the awaitables are done. However, to update a progress bar, we need to get results as they are done.

Fortunately, there is an `asyncio` equivalent of the `as_completed` generator function we used in the thread pool example with the progress bar (Example 20-16).

Example 21-6 shows the top of the *flags2_asyncio.py* script where the `get_flag` and `download_one` coroutines are defined. Example 21-7 lists the rest of the source, with `supervisor` and `download_many`. This script is longer than *flags_asyncio.py* because of error handling.

Example 21-6. flags2_asyncio.py: top portion of the script; remaining code is in Example 21-7

```
import asyncio
from collections import Counter
from http import HTTPStatus
from pathlib import Path

import httpx
import tqdm  # type: ignore

from flags2_common import main, DownloadStatus, save_flag

# low concurrency default to avoid errors from remote site,
# such as 503 - Service Temporarily Unavailable
DEFAULT_CONCUR_REQ = 5
MAX_CONCUR_REQ = 1000

async def get_flag(client: httpx.AsyncClient,  ❶
                   base_url: str,
                   cc: str) -> bytes:
    url = f'{base_url}/{cc}/{cc}.gif'.lower()
```

```
        resp = await client.get(url, timeout=3.1, follow_redirects=True)  ❷
        resp.raise_for_status()
        return resp.content

async def download_one(client: httpx.AsyncClient,
                       cc: str,
                       base_url: str,
                       semaphore: asyncio.Semaphore,
                       verbose: bool) -> DownloadStatus:
    try:
        async with semaphore:  ❸
            image = await get_flag(client, base_url, cc)
    except httpx.HTTPStatusError as exc:  ❹
        res = exc.response
        if res.status_code == HTTPStatus.NOT_FOUND:
            status = DownloadStatus.NOT_FOUND
            msg = f'not found: {res.url}'
        else:
            raise
    else:
        await asyncio.to_thread(save_flag, image, f'{cc}.gif')  ❺
        status = DownloadStatus.OK
        msg = 'OK'
    if verbose and msg:
        print(cc, msg)
    return status
```

❶ get_flag is very similar to the sequential version in Example 20-14. First difference: it requires the client parameter.

❷ Second and third differences: .get is an AsyncClient method, and it's a coroutine, so we need to await it.

❸ Use the semaphore as an asynchronous context manager so that the program as a
whole is not blocked; only this coroutine is suspended when the semaphore
counter is zero. More about this in "Python's Semaphores" on page 795.

❹ The error handling logic is the same as in download_one, from Example 20-14.

❺ Saving the image is an I/O operation. To avoid blocking the event loop, run
save_flag in a thread.

All network I/O is done with coroutines in *asyncio*, but not file I/O. However, file
I/O is also "blocking"—in the sense that reading/writing files takes thousands of
times longer (*https://fpy.li/21-15*) than reading/writing to RAM. If you're using
Network-Attached Storage (*https://fpy.li/21-16*), it may even involve network I/O
under the covers.

Since Python 3.9, the `asyncio.to_thread` coroutine makes it easy to delegate file I/O to a thread pool provided by *asyncio*. If you need to support Python 3.7 or 3.8, "Delegating Tasks to Executors" on page 801 shows how to add a couple of lines to do it. But first, let's finish our study of the HTTP client code.

Throttling Requests with a Semaphore

Network clients like the ones we are studying should be *throttled* (i.e., limited) to avoid pounding the server with too many concurrent requests.

A *semaphore* (*https://fpy.li/21-17*) is a synchronization primitive, more flexible than a lock. A semaphore can be held by multiple coroutines, with a configurable maximum number. This makes it ideal to throttle the number of active concurrent coroutines. "Python's Semaphores" on page 795 has more information.

In *flags2_threadpool.py* (Example 20-16), the throttling was done by instantiating the `ThreadPoolExecutor` with the required `max_workers` argument set to `concur_req` in the `download_many` function. In *flags2_asyncio.py*, an `asyncio.Semaphore` is created by the `supervisor` function (shown in Example 21-7) and passed as the `semaphore` argument to `download_one` in Example 21-6.

Python's Semaphores

Computer scientist Edsger W. Dijkstra invented the semaphore (*https://fpy.li/21-17*) in the early 1960s. It's a simple idea, but it's so flexible that most other synchronization objects—such as locks and barriers—can be built on top of semaphores. There are three `Semaphore` classes in Python's standard library: one in `threading`, another in `multiprocessing`, and a third one in `asyncio`. Here we'll describe the latter.

An `asyncio.Semaphore` has an internal counter that is decremented whenever we `await` on the `.acquire()` coroutine method, and incremented when we call the `.release()` method—which is not a coroutine because it never blocks. The initial value of the counter is set when the `Semaphore` is instantiated:

```
semaphore = asyncio.Semaphore(concur_req)
```

Awaiting on `.acquire()` causes no delay when the counter is greater than zero, but if the counter is zero, `.acquire()` suspends the awaiting coroutine until some other coroutine calls `.release()` on the same `Semaphore`, thus incrementing the counter. Instead of using those methods directly, it's safer to use the `semaphore` as an asynchronous context manager, as I did in Example 21-6, function `download_one`:

```
async with semaphore:
    image = await get_flag(client, base_url, cc)
```

The Semaphore.__aenter__ coroutine method awaits for .acquire(), and its __aexit__ coroutine method calls .release(). That snippet guarantees that no more than concur_req instances of get_flags coroutines will be active at any time.

Each of the Semaphore classes in the standard library has a BoundedSemaphore subclass that enforces an additional constraint: the internal counter can never become larger than the initial value when there are more .release() than .acquire() operations.[7]

Now let's take a look at the rest of the script in Example 21-7.

Example 21-7. flags2_asyncio.py: script continued from Example 21-6

```python
async def supervisor(cc_list: list[str],
                     base_url: str,
                     verbose: bool,
                     concur_req: int) -> Counter[DownloadStatus]:  ❶
    counter: Counter[DownloadStatus] = Counter()
    semaphore = asyncio.Semaphore(concur_req)  ❷
    async with httpx.AsyncClient() as client:
        to_do = [download_one(client, cc, base_url, semaphore, verbose)
                 for cc in sorted(cc_list)]  ❸
        to_do_iter = asyncio.as_completed(to_do)  ❹
        if not verbose:
            to_do_iter = tqdm.tqdm(to_do_iter, total=len(cc_list))  ❺
        error: httpx.HTTPError | None = None  ❻
        for coro in to_do_iter:  ❼
            try:
                status = await coro  ❽
            except httpx.HTTPStatusError as exc:
                error_msg = 'HTTP error {resp.status_code} - {resp.reason_phrase}'
                error_msg = error_msg.format(resp=exc.response)
                error = exc  ❾
            except httpx.RequestError as exc:
                error_msg = f'{exc} {type(exc)}'.strip()
                error = exc  ❿
            except KeyboardInterrupt:
                break

            if error:
                status = DownloadStatus.ERROR  ⓫
                if verbose:
                    url = str(error.request.url)  ⓬
                    cc = Path(url).stem.upper()  ⓭
```

---

7 Thanks to Guto Maia who noted that the concept of a semaphore was not explained when he read the first edition draft for this chapter.

```
            print(f'{cc} error: {error_msg}')
        counter[status] += 1

    return counter

def download_many(cc_list: list[str],
                  base_url: str,
                  verbose: bool,
                  concur_req: int) -> Counter[DownloadStatus]:
    coro = supervisor(cc_list, base_url, verbose, concur_req)
    counts = asyncio.run(coro)  ⓮

    return counts

if __name__ == '__main__':
    main(download_many, DEFAULT_CONCUR_REQ, MAX_CONCUR_REQ)
```

❶ supervisor takes the same arguments as the download_many function, but it cannot be invoked directly from main because it's a coroutine and not a plain function like download_many.

❷ Create an asyncio.Semaphore that will not allow more than concur_req active
coroutines among those using this semaphore. The value of concur_req is computed by the main function from *flags2_common.py*, based on command-line
options and constants set in each example.

❸ Create a list of coroutine objects, one per call to the download_one coroutine.

❹ Get an iterator that will return coroutine objects as they are done. I did not place
this call to as_completed directly in the for loop below because I may need to
wrap it with the tqdm iterator for the progress bar, depending on the user's choice
for verbosity.

❺ Wrap the as_completed iterator with the tqdm generator function to display
progress.

❻ Declare and initialize error with None; this variable will be used to hold an
exception beyond the try/except statement, if one is raised.

❼ Iterate over the completed coroutine objects; this loop is similar to the one in
download_many in Example 20-16.

❽ await on the coroutine to get its result. This will not block because as_comple
ted only produces coroutines that are done.

❾ This assignment is necessary because the `exc` variable scope is limited to this `except` clause, but I need to preserve its value for later.

❿ Same as before.

⓫ If there was an error, set the `status`.

⓬ In verbose mode, extract the URL from the exception that was raised…

⓭ …and extract the name of the file to display the country code next.

⓮ `download_many` instantiates the `supervisor` coroutine object and passes it to the event loop with `asyncio.run`, collecting the counter `supervisor` returns when the event loop ends.

In Example 21-7, we could not use the mapping of futures to country codes we saw in Example 20-16, because the awaitables returned by `asyncio.as_completed` are the same awaitables we pass into the `as_completed` call. Internally, the *asyncio* machinery may replace the awaitables we provide with others that will, in the end, produce the same results.[8]

Because I could not use the awaitables as keys to retrieve the country code from a `dict` in case of failure, I had to extract the country code from the exception. To do that, I kept the exception in the `error` variable to retrieve outside of the `try/except` statement. Python is not a block-scoped language: statements such as loops and `try/except` don't create a local scope in the blocks they manage. But if an `except` clause binds an exception to a variable, like the `exc` variables we just saw—that binding only exists within the block inside that particular `except` clause.

This wraps up the discussion of an *asyncio* example functionally equivalent to the *flags2_threadpool.py* we saw earlier.

The next example demonstrates the simple pattern of executing one asynchronous task after another using coroutines. This deserves our attention because anyone with previous experience with JavaScript knows that running one asynchronous function after the other was the reason for the nested coding pattern known as *pyramid of*

---

8 A detailed discussion about this can be found in a thread I started in the python-tulip group, titled "Which other futures may come out of asyncio.as_completed?" (*https://fpy.li/21-19*). Guido responds, and gives insight on the implementation of `as_completed`, as well as the close relationship between futures and coroutines in *asyncio*.

doom (https://fpy.li/21-20). The `await` keyword makes that curse go away. That's why `await` is now part of Python and JavaScript.

Making Multiple Requests for Each Download

Suppose you want to save each country flag with the name of the country and the country code, instead of just the country code. Now you need to make two HTTP requests per flag: one to get the flag image itself, the other to get the *metadata.json* file in the same directory as the image—that's where the name of the country is recorded.

Coordinating multiple requests in the same task is easy in the threaded script: just make one request then the other, blocking the thread twice, and keeping both pieces of data (country code and name) in local variables, ready to use when saving the files. If you needed to do the same in an asynchronous script with callbacks, you needed nested functions so that the country code and name were available in their closures until you could save the file, because each callback runs in a different local scope. The `await` keyword provides relief from that, allowing you to drive the asynchronous requests one after the other, sharing the local scope of the driving coroutine.

> If you are doing asynchronous application programming in modern Python with lots of callbacks, you are probably applying old patterns that don't make sense in modern Python. That is justified if you are writing a library that interfaces with legacy or low-level code that does not support coroutines. Anyway, the StackOverflow Q&A, "What is the use case for future.add_done_callback()?" (*https://fpy.li/21-21*) explains why callbacks are needed in low-level code, but are not very useful in Python application-level code these days.

The third variation of the `asyncio` flag downloading script has a few changes:

`get_country`
> This new coroutine fetches the *metadata.json* file for the country code, and gets the name of the country from it.

`download_one`
> This coroutine now uses `await` to delegate to `get_flag` and the new `get_country` coroutine, using the result of the latter to build the name of the file to save.

Let's start with the code for `get_country` (Example 21-8). Note that it is very similar to `get_flag` from Example 21-6.

Example 21-8. flags3_asyncio.py: get_country coroutine

```python
async def get_country(client: httpx.AsyncClient,
                      base_url: str,
                      cc: str) -> str:        ❶
    url = f'{base_url}/{cc}/metadata.json'.lower()
    resp = await client.get(url, timeout=3.1, follow_redirects=True)
    resp.raise_for_status()
    metadata = resp.json()    ❷
    return metadata['country']    ❸
```

❶   This coroutine returns a string with the country name—if all goes well.

❷   metadata will get a Python dict built from the JSON contents of the response.

❸   Return the country name.

Now let's see the modified download_one in Example 21-9, which has only a few lines changed from the same coroutine in Example 21-6.

Example 21-9. flags3_asyncio.py: download_one coroutine

```python
async def download_one(client: httpx.AsyncClient,
                       cc: str,
                       base_url: str,
                       semaphore: asyncio.Semaphore,
                       verbose: bool) -> DownloadStatus:
    try:
        async with semaphore:    ❶
            image = await get_flag(client, base_url, cc)
        async with semaphore:    ❷
            country = await get_country(client, base_url, cc)
    except httpx.HTTPStatusError as exc:
        res = exc.response
        if res.status_code == HTTPStatus.NOT_FOUND:
            status = DownloadStatus.NOT_FOUND
            msg = f'not found: {res.url}'
        else:
            raise
    else:
        filename = country.replace(' ', '_')    ❸
        await asyncio.to_thread(save_flag, image, f'{filename}.gif')
        status = DownloadStatus.OK
        msg = 'OK'
    if verbose and msg:
        print(cc, msg)
    return status
```

❶   Hold the semaphore to await for get_flag...

---

❷ …and again for `get_country`.

❸ Use the country name to create a filename. As a command-line user, I don't like to see spaces in filenames.

Much better than nested callbacks!

I put the calls to `get_flag` and `get_country` in separate `with` blocks controlled by the `semaphore` because it's good practice to hold semaphores and locks for the shortest possible time.

I could schedule both `get_flag` and `get_country` in parallel using `asyncio.gather`, but if `get_flag` raises an exception, there is no image to save, so it's pointless to run `get_country`. But there are cases where it makes sense to use `asyncio.gather` to hit several APIs at the same time instead of waiting for one response before making the next request.

In *flags3_asyncio.py*, the `await` syntax appears six times, and `async with` three times. Hopefully, you should be getting the hang of asynchronous programming in Python. One challenge is to know when you have to use `await` and when you can't use it. The answer in principle is easy: you `await` coroutines and other awaitables, such as `asyncio.Task` instances. But some APIs are tricky, mixing coroutines and plain functions in seemingly arbitrary ways, like the `StreamWriter` class we'll use in Example 21-14.

Example 21-9 wrapped up the *flags* set of examples. Now let's discuss the use of thread or process executors in asynchronous programming.

Delegating Tasks to Executors

One important advantage of Node.js over Python for asynchronous programming is the Node.js standard library, which provides async APIs for all I/O—not just for network I/O. In Python, if you're not careful, file I/O can seriously degrade the performance of asynchronous applications, because reading and writing to storage in the main thread blocks the event loop.

In the `download_one` coroutine of Example 21-6, I used this line to save the downloaded image to disk:

```
await asyncio.to_thread(save_flag, image, f'{cc}.gif')
```

As mentioned before, the `asyncio.to_thread` was added in Python 3.9. If you need to support 3.7 or 3.8, then replace that single line with the lines in Example 21-10.

Example 21-10. Lines to use instead of `await asyncio.to_thread`

```
loop = asyncio.get_running_loop()          ❶
loop.run_in_executor(None, save_flag,      ❷
                     image, f'{cc}.gif')   ❸
```

❶ Get a reference to the event loop.

❷ The first argument is the executor to use; passing None selects the default Thread
    PoolExecutor that is always available in the asyncio event loop.

❸ You can pass positional arguments to the function to run, but if you need to pass
    keyword arguments, then you need to resort to functtool.partial, as described
    in the run_in_executor documentation (*https://fpy.li/21-22*).

The newer asyncio.to_thread function is easier to use and more flexible, as it also
accepts keyword arguments.

The implementation of asyncio itself uses run_in_executor under the hood in a few
places. For example, the loop.getaddrinfo(…) coroutine we saw in Example 21-1
is implemented by calling the getaddrinfo function from the socket module—
which is a blocking function that may take seconds to return, as it depends on DNS
resolution.

A common pattern in asynchronous APIs is to wrap blocking calls that are implementation details in coroutines using run_in_executor internally. That way, you
provide a consistent interface of coroutines to be driven with await, and hide the
threads you need to use for pragmatic reasons. The Motor (*https://fpy.li/21-23*) asynchronous driver for MongoDB has an API compatible with async/await that is really
a façade around a threaded core that talks to the database server. A. Jesse Jiryu Davis,
the lead developer of Motor, explains his reasoning in "Response to 'Asynchronous
Python and Databases'" (*https://fpy.li/21-24*). Spoiler: Davis discovered that a thread
pool was more performant in the particular use case of a database driver—despite the
myth that asynchronous approaches are always faster than threads for network I/O.

The main reason to pass an explicit Executor to loop.run_in_executor is to employ
a ProcessPoolExecutor if the function to execute is CPU intensive, so that it runs in
a different Python process, avoiding contention for the GIL. Because of the high
start-up cost, it would be better to start the ProcessPoolExecutor in the supervisor,
and pass it to the coroutines that need to use it.

Caleb Hattingh—the author of *Using Asyncio in Python* (O' Reilly)—is one of the
tech reviewers of this book and suggested I add the following warning about executors and *asyncio*.

Caleb's Warning about run_in_executors

Using `run_in_executor` can produce hard-to-debug problems since cancellation doesn't work the way one might expect. Coroutines that use executors give merely the pretense of cancellation: the underlying thread (if it's a `ThreadPoolExecutor`) has no cancellation mechanism. For example, a long-lived thread that is created inside a `run_in_executor` call may prevent your *asyncio* program from shutting down cleanly: `asyncio.run` will wait for the executor to fully shut down before returning, and it will wait forever if the executor jobs don't stop somehow on their own. My greybeard inclination is to want that function to be named `run_in_executor_uncancellable`.

We'll now go from client scripts to writing servers with `asyncio`.

Writing asyncio Servers

The classic toy example of a TCP server is an echo server (*https://fpy.li/21-25*). We'll build slightly more interesting toys: server-side Unicode character search utilities, first using HTTP with *FastAPI*, then using plain TCP with `asyncio` only.

These servers let users query for Unicode characters based on words in their standard names from the `unicodedata` module we discussed in "The Unicode Database" on page 151. Figure 21-2 shows a session with *web_mojifinder.py*, the first server we'll build.

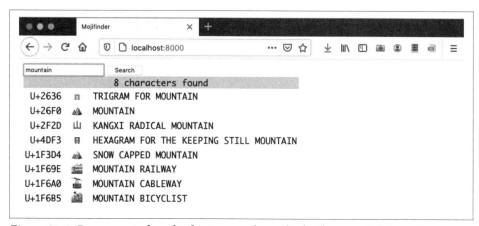

Figure 21-2. Browser window displaying search results for "mountain" from the web_mojifinder.py service.

The Unicode search logic in these examples is in the `InvertedIndex` class in the *char-index.py* module in the *Fluent Python* code repository (*https://fpy.li/code*). There's nothing concurrent in that small module, so I'll only give a brief overview in the optional box that follows. You can skip to the HTTP server implementation in "A FastAPI Web Service" on page 805.

Meet the Inverted Index

An inverted index usually maps words to documents in which they occur. In the *mojifinder* examples, each "document" is one Unicode character. The `charin dex.InvertedIndex` class indexes each word that appears in each character name in the Unicode database, and creates an inverted index stored in a `defaultdict`. For example, to index character U+0037—DIGIT SEVEN—the `InvertedIndex` initializer appends the character `'7'` to the entries under the keys `'DIGIT'` and `'SEVEN'`. After indexing the Unicode 13.0.0 data bundled with Python 3.9.1, `'DIGIT'` maps to 868 characters, and `'SEVEN'` maps to 143, including U+1F556—CLOCK FACE SEVEN OCLOCK and U+2790—DINGBAT NEGATIVE CIRCLED SANS-SERIF DIGIT SEVEN (which appears in many code listings in this book).

See Figure 21-3 for a demonstration using the entries for `'CAT'` and `'FACE'`.[9]

```
>>> from charindex import InvertedIndex
>>> idx.entries['CAT']
{'😺', '😸', '찰', '🐱', '😹', '😻', '😼', '😽', '😾', '😿', '🐈', '0 ', '😺', '😺'}
>>> len(idx.entries['FACE'])
171
>>> idx.entries['FACE'] & idx.entries['CAT']
{'😺', '😸', '😹', '😻', '😼', '😽', '😾', '😿', '😺', '😺'}
>>> idx.search('cat face')
{'😺', '😸', '😹', '😻', '😼', '😽', '😾', '😿', '😺', '😺'}
>>>
```

Figure 21-3. Python console exploring `InvertedIndex` attribute `entries` and `search` method.

The `InvertedIndex.search` method breaks the query into words, and returns the intersection of the entries for each word. That's why searching for "face" finds 171 results, "cat" finds 14, but "cat face" only 10.

That's the beautiful idea behind an inverted index: a fundamental building block in information retrieval—the theory behind search engines. See the English Wikipedia article "Inverted Index" (*https://fpy.li/21-27*) to learn more.

---

9 The boxed question mark in the screen shot is not a defect of the book or ebook you are reading. It's the U +101EC—PHAISTOS DISC SIGN CAT character, which is missing from the font in the terminal I used. The Phaistos disc (*https://fpy.li/21-26*) is an ancient artifact inscribed with pictograms, discovered in the island of Crete.

A FastAPI Web Service

I wrote the next example—*web_mojifinder.py*—using *FastAPI* (*https://fpy.li/21-28*): one of the Python ASGI Web frameworks mentioned in "ASGI—Asynchronous Server Gateway Interface" on page 736. Figure 21-2 is a screenshot of the frontend. It's a super simple SPA (Single Page Application): after the initial HTML download, the UI is updated by client-side JavaScript communicating with the server.

FastAPI is designed to implement backends for SPA and mobile apps, which mostly consist of web API end points returning JSON responses instead of server-rendered HTML. *FastAPI* leverages decorators, type hints, and code introspection to eliminate a lot of the boilerplate code for web APIs, and also automatically publishes interactive OpenAPI—a.k.a. Swagger (*https://fpy.li/21-29*)—documentation for the APIs we create. Figure 21-4 shows the autogenerated /docs page for *web_mojifinder.py*.

Figure 21-4. Autogenerated OpenAPI schema for the /search endpoint.

Example 21-11 is the code for *web_mojifinder.py*, but that's just the backend code. When you hit the root URL /, the server sends the *form.html* file, which has 81 lines of code, including 54 lines of JavaScript to communicate with the server and fill a table with the results. If you're interested in reading plain framework-less JavaScript, please find *21-async/mojifinder/static/form.html* in the *Fluent Python* code repository (*https://fpy.li/code*).

To run *web_mojifinder.py*, you need to install two packages and their dependencies: *FastAPI* and *uvicorn*.[10] This is the command to run Example 21-11 with *uvicorn* in development mode:

```
$ uvicorn web_mojifinder:app --reload
```

The parameters are:

web_mojifinder:app
> The package name, a colon, and the name of the ASGI application defined in it—app is the conventional name.

--reload
> Make *uvicorn* monitor changes to application source files and automatically reload them. Useful only during development.

Now let's study the source code for *web_mojifinder.py*.

Example 21-11. web_mojifinder.py: complete source

```
from pathlib import Path
from unicodedata import name

from fastapi import FastAPI
from fastapi.responses import HTMLResponse
from pydantic import BaseModel

from charindex import InvertedIndex

STATIC_PATH = Path(__file__).parent.absolute() / 'static'  ❶

app = FastAPI(  ❷
    title='Mojifinder Web',
    description='Search for Unicode characters by name.',
)

class CharName(BaseModel):  ❸
    char: str
    name: str

def init(app):  ❹
    app.state.index = InvertedIndex()
    app.state.form = (STATIC_PATH / 'form.html').read_text()

init(app)  ❺
```

---

10 Instead of *uvicorn*, you may use another ASGI server, such as *hypercorn* or *Daphne*. See the official ASGI documentation page about implementations (*https://fpy.li/21-30*) for more information.

```
@app.get('/search', response_model=list[CharName])  ❻
async def search(q: str):  ❼
    chars = sorted(app.state.index.search(q))
    return ({'char': c, 'name': name(c)} for c in chars)  ❽

@app.get('/', response_class=HTMLResponse, include_in_schema=False)
def form():  ❾
    return app.state.form

# no main funcion  ❿
```

❶ Unrelated to the theme of this chapter, but worth noting: the elegant use of the overloaded / operator by pathlib.[11]

❷ This line defines the ASGI app. It could be as simple as app = FastAPI(). The parameters shown are metadata for the autogenerated documentation.

❸ A *pydantic* schema for a JSON response with char and name fields.[12]

❹ Build the index and load the static HTML form, attaching both to the app.state for later use.

❺ Run init when this module is loaded by the ASGI server.

❻ Route for the /search endpoint; response_model uses that CharName *pydantic* model to describe the response format.

❼ *FastAPI* assumes that any parameters that appear in the function or coroutine signature that are not in the route path will be passed in the HTTP query string, e.g., /search?q=cat. Since q has no default, *FastAPI* will return a 422 (Unprocessable Entity) status if q is missing from the query string.

❽ Returning an iterable of dicts compatible with the response_model schema allows *FastAPI* to build the JSON response according to the response_model in the @app.get decorator.

❾ Regular functions (i.e., non-async) can also be used to produce responses.

❿ This module has no main function. It is loaded and driven by the ASGI server— *uvicorn* in this example.

---

11 Thanks to tech reviewer Miroslav Šedivý for highlighting good places to use pathlib in code examples.

12 As mentioned in Chapter 8, *pydantic* (*https://fpy.li/21-31*) enforces type hints at runtime, for data validation.

Example 21-11 has no direct calls to asyncio. *FastAPI* is built on the *Starlette* ASGI toolkit, which in turn uses asyncio.

Also note that the body of search doesn't use await, async with, or async for, therefore it could be a plain function. I defined search as a coroutine just to show that *FastAPI* knows how to handle it. In a real app, most endpoints will query databases or hit other remote servers, so it is a critical advantage of *FastAPI*—and ASGI frameworks in general—to support coroutines that can take advantage of asynchronous libraries for network I/O.

> The init and form functions I wrote to load and serve the static HTML form are a hack to make the example short and easy to run. The recommended best practice is to have a proxy/load-balancer in front of the ASGI server to handle all static assets, and also use a CDN (Content Delivery Network) when possible. One such proxy/load-balancer is *Traefik* (*https://fpy.li/21-32*), a self-described "edge router" that "receives requests on behalf of your system and finds out which components are responsible for handling them." *FastAPI* has project generation (*https://fpy.li/21-33*) scripts that prepare your code to do that.

The typing enthusiast may have noticed that there are no return type hints in search and form. Instead, *FastAPI* relies on the response_model= keyword argument in the route decorators. The "Response Model" (*https://fpy.li/21-34*) page in the *FastAPI* documentation explains:

> The response model is declared in this parameter instead of as a function return type annotation, because the path function may not actually return that response model but rather return a dict, database object or some other model, and then use the response_model to perform the field limiting and serialization.

For example, in search, I returned a generator of dict items, not a list of CharName objects, but that's good enough for *FastAPI* and *pydantic* to validate my data and build the appropriate JSON response compatible with response_model=list[Char Name].

We'll now focus on the *tcp_mojifinder.py* script that is answering the queries in Figure 21-5.

An asyncio TCP Server

The *tcp_mojifinder.py* program uses plain TCP to communicate with a client like Telnet or Netcat, so I could write it using asyncio without external dependencies—and without reinventing HTTP. Figure 21-5 shows text-based UI.

```
 ● ● ●                    ⌂ luciano — telnet localhost 2323 — 83×30
TW–LR–MBP:~ luciano$ telnet localhost 2323
Trying 127.0.0.1...
Connected to localhost.
Escape character is '^]'.
?> fire
U+2632  ☲     TRIGRAM FOR FIRE
U+2EA3  ⺣     CJK RADICAL FIRE
U+2F55  火     KANGXI RADICAL FIRE
U+322B  ㈫     PARENTHESIZED IDEOGRAPH FIRE
U+328B  ㊋     CIRCLED IDEOGRAPH FIRE
U+4DDD  ䷝     HEXAGRAM FOR THE CLINGING FIRE
U+1F525 🔥     FIRE
U+1F692 🚒     FIRE ENGINE
U+1F6F1 🛱     ONCOMING FIRE ENGINE
U+1F702 🜂     ALCHEMICAL SYMBOL FOR FIRE
U+1F9EF 🧯     FIRE EXTINGUISHER
 ─────────────────────────────────────────────────── 11 found
```

Figure 21-5. Telnet session with the tcp_mojifinder.py server: querying for "fire."

This program is twice as long as *web_mojifinder.py*, so I split the presentation into three parts: Example 21-12, Example 21-14, and Example 21-15. The top of *tcp_moji-finder.py*—including the import statements—is in Example 21-14, but I will start by describing the supervisor coroutine and the main function that drives the program.

Example 21-12. tcp_mojifinder.py: a simple TCP server; continues in Example 21-14

```python
async def supervisor(index: InvertedIndex, host: str, port: int) -> None:
    server = await asyncio.start_server(        ❶
        functools.partial(finder, index),       ❷
        host, port)                             ❸

    socket_list = cast(tuple[TransportSocket, ...], server.sockets)  ❹
    addr = socket_list[0].getsockname()
    print(f'Serving on {addr}. Hit CTRL-C to stop.')  ❺
    await server.serve_forever()  ❻

def main(host: str = '127.0.0.1', port_arg: str = '2323'):
    port = int(port_arg)
    print('Building index.')
    index = InvertedIndex()                     ❼
    try:
        asyncio.run(supervisor(index, host, port))  ❽
    except KeyboardInterrupt:                    ❾
        print('\nServer shut down.')

if __name__ == '__main__':
    main(*sys.argv[1:])
```

❶ This `await` quickly gets an instance of `asyncio.Server`, a TCP socket server. By default, `start_server` creates and starts the server, so it's ready to receive connections.

❷ The first argument to `start_server` is `client_connected_cb`, a callback to run when a new client connection starts. The callback can be a function or a coroutine, but it must accept exactly two arguments: an `asyncio.StreamReader` and an `asyncio.StreamWriter`. However, my `finder` coroutine also needs to get an `index`, so I used `functools.partial` to bind that parameter and obtain a callable that takes the reader and writer. Adapting user functions to callback APIs is the most common use case for `functools.partial`.

❸ `host` and `port` are the second and third arguments to `start_server`. See the full signature in the `asyncio` documentation (*https://fpy.li/21-35*).

❹ This `cast` is needed because *typeshed* has an outdated type hint for the `sockets` property of the `Server` class—as of May 2021. See Issue #5535 on *typeshed* (*https://fpy.li/21-36*).[13]

❺ Display the address and port of the first socket of the server.

❻ Although `start_server` already started the server as a concurrent task, I need to `await` on the `server_forever` method so that my `supervisor` is suspended here. Without this line, `supervisor` would return immediately, ending the loop started with `asyncio.run(supervisor(…))`, and exiting the program. The documentation for `Server.serve_forever` (*https://fpy.li/21-37*) says: "This method can be called if the server is already accepting connections."

❼ Build the inverted index.[14]

❽ Start the event loop running `supervisor`.

❾ Catch the `KeyboardInterrupt` to avoid a distracting traceback when I stop the server with Ctrl-C on the terminal running it.

---

13 Issue #5535 is closed as of October 2021, but Mypy did not have a new release since then, so the error persists.

14 Tech reviewer Leonardo Rochael pointed out that building the index could be delegated to another thread using `loop.run_with_executor()` in the `supervisor` coroutine, so the server would be ready to take requests immediately while the index is built. That's true, but querying the index is the only thing this server does, so it would not be a big win in this example.

You may find it easier to understand how control flows in *tcp_mojifinder.py* if you study the output it generates on the server console, listed in Example 21-13.

Example 21-13. tcp_mojifinder.py: this is the server side of the session depicted in Figure 21-5

```
$ python3 tcp_mojifinder.py
Building index.  ❶
Serving on ('127.0.0.1', 2323). Hit Ctrl-C to stop.  ❷
 From ('127.0.0.1', 58192): 'cat face'  ❸
   To ('127.0.0.1', 58192): 10 results.
 From ('127.0.0.1', 58192): 'fire'      ❹
   To ('127.0.0.1', 58192): 11 results.
 From ('127.0.0.1', 58192): '\x00'      ❺
Close ('127.0.0.1', 58192).             ❻
^C ❼
Server shut down.  ❽
$
```

❶ Output by main. Before the next line appears, I see a 0.6s delay on my machine while the index is built.

❷ Output by supervisor.

❸ First iteration of a while loop in finder. The TCP/IP stack assigned port 58192 to my Telnet client. If you connect several clients to the server, you'll see their various ports in the output.

❹ Second iteration of the while loop in finder.

❺ I hit Ctrl-C on the client terminal; the while loop in finder exits.

❻ The finder coroutine displays this message then exits. Meanwhile the server is still running, ready to service another client.

❼ I hit Ctrl-C on the server terminal; server.serve_forever is cancelled, ending supervisor and the event loop.

❽ Output by main.

After main builds the index and starts the event loop, supervisor quickly displays the Serving on… message and is suspended at the await server.serve_forever() line. At that point, control flows into the event loop and stays there, occasionally coming back to the finder coroutine, which yields control back to the event loop whenever it needs to wait for the network to send or receive data.

While the event loop is alive, a new instance of the finder coroutine will be started for each client that connects to the server. In this way, many clients can be handled concurrently by this simple server. This continues until a KeyboardInterrupt occurs on the server or its process is killed by the OS.

Now let's see the top of *tcp_mojifinder.py*, with the finder coroutine.

Example 21-14. tcp_mojifinder.py: continued from Example 21-12

```python
import asyncio
import functools
import sys
from asyncio.trsock import TransportSocket
from typing import cast

from charindex import InvertedIndex, format_results  ❶

CRLF = b'\r\n'
PROMPT = b'?> '

async def finder(index: InvertedIndex,               ❷
                 reader: asyncio.StreamReader,
                 writer: asyncio.StreamWriter) -> None:
    client = writer.get_extra_info('peername')  ❸
    while True:  ❹
        writer.write(PROMPT)  # can't await!  ❺
        await writer.drain()  # must await!  ❻
        data = await reader.readline()  ❼
        if not data:  ❽
            break
        try:
            query = data.decode().strip()  ❾
        except UnicodeDecodeError:  ❿
            query = '\x00'
        print(f' From {client}: {query!r}')  ⓫
        if query:
            if ord(query[:1]) < 32:  ⓬
                break
            results = await search(query, index, writer)  ⓭
            print(f'   To {client}: {results} results.')  ⓮

    writer.close()  ⓯
    await writer.wait_closed()  ⓰
    print(f'Close {client}.')  ⓱
```

❶ format_results is useful to display the results of InvertedIndex.search in a text-based UI such as the command line or a Telnet session.

❷ To pass finder to asyncio.start_server, I wrapped it with functools.par tial, because the server expects a coroutine or function that takes only the reader and writer arguments.

❸ Get the remote client address to which the socket is connected.

❹ This loop handles a dialog that lasts until a control character is received from the client.

❺ The StreamWriter.write method is not a coroutine, just a plain function; this line sends the ?> prompt.

❻ StreamWriter.drain flushes the writer buffer; it is a coroutine, so it must be driven with await.

❼ StreamWriter.readline is a coroutine that returns bytes.

❽ If no bytes were received, the client closed the connection, so exit the loop.

❾ Decode the bytes to str, using the default UTF-8 encoding.

❿ A UnicodeDecodeError may happen when the user hits Ctrl-C and the Telnet client sends control bytes; if that happens, replace the query with a null character, for simplicity.

⓫ Log the query to the server console.

⓬ Exit the loop if a control or null character was received.

⓭ Do the actual search; code is presented next.

⓮ Log the response to the server console.

⓯ Close the StreamWriter.

⓰ Wait for the StreamWriter to close. This is recommended in the .close() method documentation (*https://fpy.li/21-38*).

⓱ Log the end of this client's session to the server console.

The last piece of this example is the search coroutine, shown in Example 21-15.

Example 21-15. tcp_mojifinder.py: search coroutine

```
async def search(query: str,  ❶
                 index: InvertedIndex,
                 writer: asyncio.StreamWriter) -> int:
    chars = index.search(query)  ❷
    lines = (line.encode() + CRLF for line  ❸
                in format_results(chars))
    writer.writelines(lines)  ❹
    await writer.drain()       ❺
    status_line = f'{"-" * 66} {len(chars)} found'  ❻
    writer.write(status_line.encode() + CRLF)
    await writer.drain()
    return len(chars)
```

❶ search must be a coroutine because it writes to a StreamWriter and must use its .drain() coroutine method.

❷ Query the inverted index.

❸ This generator expression will yield byte strings encoded in UTF-8 with the Unicode codepoint, the actual character, its name, and a CRLF sequence—e.g., b'U+0039\t9\tDIGIT NINE\r\n').

❹ Send the lines. Surprisingly, writer.writelines is not a coroutine.

❺ But writer.drain() is a coroutine. Don't forget the await!

❻ Build a status line, then send it.

Note that all network I/O in *tcp_mojifinder.py* is in bytes; we need to decode the bytes received from the network, and encode strings before sending them out. In Python 3, the default encoding is UTF-8, and that's what I used implicitly in all encode and decode calls in this example.

 Note that some of the I/O methods are coroutines and must be driven with await, while others are simple functions. For example, StreamWriter.write is a plain function, because it writes to a buffer. On the other hand, StreamWriter.drain—which flushes the buffer and performs the network I/O—is a coroutine, as is StreamReader.readline—but not StreamWriter.writelines! While I was writing the first edition of this book, the asyncio API docs were improved by clearly labeling coroutines as such (*https://fpy.li/21-39*).

The *tcp_mojifinder.py* code leverages the high-level `asyncio` Streams API (*https://fpy.li/21-40*) that provides a ready-to-use server so you only need to implement a handler function, which can be a plain callback or a coroutine. There is also a lower-level Transports and Protocols API (*https://fpy.li/21-41*), inspired by the transport and protocols abstractions in the *Twisted* framework. Refer to the `asyncio` documentation for more information, including TCP and UDP echo servers and clients (*https://fpy.li/21-42*) implemented with that lower-level API.

Our next topic is `async for` and the objects that make it work.

Asynchronous Iteration and Asynchronous Iterables

We saw in "Asynchronous Context Managers" on page 790 how `async with` works with objects implementing the `__aenter__` and `__aexit__` methods returning awaitables—usually in the form of coroutine objects.

Similarly, `async for` works with *asynchronous iterables*: objects that implement `__aiter__`. However, `__aiter__` must be a regular method—not a coroutine method —and it must return an *asynchronous iterator*.

An asynchronous iterator provides an `__anext__` coroutine method that returns an awaitable—often a coroutine object. They are also expected to implement `__aiter__`, which usually returns `self`. This mirrors the important distinction of iterables and iterators we discussed in "Don't Make the Iterable an Iterator for Itself" on page 609.

The *aiopg* asynchronous PostgreSQL driver documentation (*https://fpy.li/21-43*) has an example that illustrates the use of `async for` to iterate over the rows of a database cursor:

```
async def go():
    pool = await aiopg.create_pool(dsn)
    async with pool.acquire() as conn:
        async with conn.cursor() as cur:
            await cur.execute("SELECT 1")
            ret = []
            async for row in cur:
                ret.append(row)
            assert ret == [(1,)]
```

In this example the query will return a single row, but in a realistic scenario you may have thousands of rows in response to a SELECT query. For large responses, the cursor will not be loaded with all the rows in a single batch. Therefore it is important that `async for row in cur:` does not block the event loop while the cursor may be waiting for additional rows. By implementing the cursor as an asynchronous iterator, *aiopg* may yield to the event loop at each `__anext__` call, and resume later when more rows arrive from PostgreSQL.

Asynchronous Generator Functions

You can implement an asynchronous iterator by writing a class with __anext__ and __aiter__, but there is a simpler way: write a function declared with async def and use yield in its body. This parallels how generator functions simplify the classic Iterator pattern.

Let's study a simple example using async for and implementing an asynchronous generator. In Example 21-1 we saw *blogdom.py*, a script that probed domain names. Now suppose we find other uses for the probe coroutine we defined there, and decide to put it into a new module—*domainlib.py*—together with a new multi_probe asynchronous generator that takes a list of domain names and yields results as they are probed.

We'll look at the implementation of *domainlib.py* soon, but first let's see how it is used with Python's new asynchronous console.

Experimenting with Python's async console

Since Python 3.8 (*https://fpy.li/21-44*), you can run the interpreter with the -m asyncio command-line option to get an "async REPL": a Python console that imports asyncio, provides a running event loop, and accepts await, async for, and async with at the top-level prompt—which otherwise are syntax errors when used outside of native coroutines.[15]

To experiment with *domainlib.py*, go to the *21-async/domains/asyncio/* directory in your local copy of the *Fluent Python* code repository (*https://fpy.li/code*). Then run:

```
$ python -m asyncio
```

You'll see the console start, similar to this:

```
asyncio REPL 3.9.1 (v3.9.1:1e5d33e9b9, Dec  7 2020, 12:10:52)
[Clang 6.0 (clang-600.0.57)] on darwin
Use "await" directly instead of "asyncio.run()".
Type "help", "copyright", "credits" or "license" for more information.
>>> import asyncio
>>>
```

Note how the header says you can use await instead of asyncio.run()—to drive coroutines and other awaitables. Also: I did not type import asyncio. The asyncio module is automatically imported and that line makes that fact clear to the user.

---

15 This is great for experimentation, like the Node.js console. Thanks to Yury Selivanov for yet another excellent contribution to asynchronous Python.

Now let's import *domainlib.py* and play with its two coroutines: `probe` and `multi_probe` (Example 21-16).

Example 21-16. Experimenting with domainlib.py after running python3 -m asyncio

```
>>> await asyncio.sleep(3, 'Rise and shine!')  ❶
'Rise and shine!'
>>> from domainlib import *
>>> await probe('python.org')  ❷
Result(domain='python.org', found=True)  ❸
>>> names = 'python.org rust-lang.org golang.org no-lang.invalid'.split()  ❹
>>> async for result in multi_probe(names):  ❺
...     print(*result, sep='\t')
...
golang.org      True  ❻
no-lang.invalid False
python.org      True
rust-lang.org   True
>>>
```

❶ Try a simple `await` to see the asynchronous console in action. Tip: `asyncio.sleep()` takes an optional second argument that is returned when you `await` it.

❷ Drive the `probe` coroutine.

❸ The `domainlib` version of `probe` returns a `Result` named tuple.

❹ Make a list of domains. The `.invalid` top-level domain is reserved for testing. DNS queries for such domains always get an NXDOMAIN response from DNS servers, meaning "that domain does not exist."[16]

❺ Iterate with `async for` over the `multi_probe` asynchronous generator to display the results.

❻ Note that the results are not in the order the domains were given to `multiprobe`. They appear as each DNS response comes back.

Example 21-16 shows that `multi_probe` is an asynchronous generator because it is compatible with `async for`. Now let's do a few more experiments, continuing from that example with Example 21-17.

---

16 See RFC 6761—Special-Use Domain Names (*https://fpy.li/21-45*).

Example 21-17. More experiments, continuing from Example 21-16

```
>>> probe('python.org')   ❶
<coroutine object probe at 0x10e313740>
>>> multi_probe(names)   ❷
<async_generator object multi_probe at 0x10e246b80>
>>> for r in multi_probe(names):   ❸
...     print(r)
...
Traceback (most recent call last):
  ...
TypeError: 'async_generator' object is not iterable
```

❶ Calling a native coroutine gives you a coroutine object.

❷ Calling an asynchronous generator gives you an async_generator object.

❸   We can't use a regular for loop with asynchronous generators because they implement __aiter__ instead of __iter__.

Asynchronous generators are driven by async for, which can be a block statement (as seen in Example 21-16), and it also appears in asynchronous comprehensions, which we'll cover soon.

Implementing an asynchronous generator

Now let's study the code for *domainlib.py*, with the multi_probe asynchronous generator (Example 21-18).

Example 21-18. domainlib.py: functions for probing domains

```
import asyncio
import socket
from collections.abc import Iterable, AsyncIterator
from typing import NamedTuple, Optional


class Result(NamedTuple):   ❶
    domain: str
    found: bool


OptionalLoop = Optional[asyncio.AbstractEventLoop]   ❷


async def probe(domain: str, loop: OptionalLoop = None) -> Result:   ❸
    if loop is None:
        loop = asyncio.get_running_loop()
    try:
```

```
        await loop.getaddrinfo(domain, None)
    except socket.gaierror:
        return Result(domain, False)
    return Result(domain, True)


async def multi_probe(domains: Iterable[str]) -> AsyncIterator[Result]:  ❹
    loop = asyncio.get_running_loop()
    coros = [probe(domain, loop) for domain in domains]  ❺
    for coro in asyncio.as_completed(coros):  ❻
        result = await coro  ❼
        yield result  ❽
```

❶ NamedTuple makes the result from probe easier to read and debug.

❷ This type alias is to avoid making the next line too long for a book listing.

❸ probe now gets an optional loop argument, to avoid repeated calls to get_run
    ning_loop when this coroutine is driven by multi_probe.

❹ An asynchronous generator function produces an asynchronous generator
    object, which can be annotated as AsyncIterator[SomeType].

❺ Build list of probe coroutine objects, each with a different domain.

❻ This is not async for because asyncio.as_completed is a classic generator.

❼ Await on the coroutine object to retrieve the result.

❽ Yield result. This line makes multi_probe an asynchronous generator.

> The for loop in Example 21-18 could be more concise:
>
> ```
> for coro in asyncio.as_completed(coros):
>     yield await coro
> ```
>
> Python parses that as yield (await coro), so it works.
>
> I thought it could be confusing to use that shortcut in the first
> asynchronous generator example in the book, so I split it into two
> lines.

Given *domainlib.py*, we can demonstrate the use of the multi_probe asynchronous
generator in *domaincheck.py*: a script that takes a domain suffix and searches for
domains made from short Python keywords.

Here is a sample output of *domaincheck.py*:

```
$ ./domaincheck.py net
FOUND           NOT FOUND
=====           =========
in.net
del.net
true.net
for.net
is.net
                none.net
try.net
                from.net
and.net
or.net
else.net
with.net
if.net
as.net
                elif.net
                pass.net
                not.net
                def.net
```

Thanks to *domainlib*, the code for *domaincheck.py* is straightforward, as seen in Example 21-19.

Example 21-19. domaincheck.py: utility for probing domains using domainlib

```python
#!/usr/bin/env python3
import asyncio
import sys
from keyword import kwlist

from domainlib import multi_probe


async def main(tld: str) -> None:
    tld = tld.strip('.')
    names = (kw for kw in kwlist if len(kw) <= 4)     ❶
    domains = (f'{name}.{tld}'.lower() for name in names)     ❷
    print('FOUND\t\tNOT FOUND')     ❸
    print('=====\t\t=========')
    async for domain, found in multi_probe(domains):     ❹
        indent = '' if found else '\t\t'     ❺
        print(f'{indent}{domain}')


if __name__ == '__main__':
    if len(sys.argv) == 2:
        asyncio.run(main(sys.argv[1]))     ❻
    else:
        print('Please provide a TLD.', f'Example: {sys.argv[0]} COM.BR')
```

❶ Generate keywords with length up to 4.

❷ Generate domain names with the given suffix as TLD.

❸ Format a header for the tabular output.

❹ Asynchronously iterate over `multi_probe(domains)`.

❺ Set `indent` to zero or two tabs to put the result in the proper column.

❻ Run the `main` coroutine with the given command-line argument.

Generators have one extra use unrelated to iteration: they can be made into context managers. This also applies to asynchronous generators.

Asynchronous generators as context managers

Writing our own asynchronous context managers is not a frequent programming task, but if you need to write one, consider using the `@asynccontextmanager` (*https://fpy.li/21-46*) decorator added to the `contextlib` module in Python 3.7. That's very similar to the `@contextmanager` decorator we studied in "Using @contextmanager" on page 668.

An interesting example combining `@asynccontextmanager` with `loop.run_in_execu tor` appears in Caleb Hattingh's book *Using Asyncio in Python* (*https://fpy.li/ hattingh*). Example 21-20 is Caleb's code—with a single change and added callouts.

Example 21-20. Example using `@asynccontextmanager` and `loop.run_in_executor`

```
from contextlib import asynccontextmanager

@asynccontextmanager
async def web_page(url):          ❶
    loop = asyncio.get_running_loop()       ❷
    data = await loop.run_in_executor(      ❸
        None, download_webpage, url)
    yield data                              ❹
    await loop.run_in_executor(None, update_stats, url)  ❺

async with web_page('google.com') as data:  ❻
    process(data)
```

❶ The decorated function must be an asynchronous generator.

❷ Minor update to Caleb's code: use the lightweight `get_running_loop` instead of `get_event_loop`.

❸ Suppose download_webpage is a blocking function using the *requests* library; we run it in a separate thread to avoid blocking the event loop.

❹ All lines before this yield expression will become the __aenter__ coroutine-method of the asynchronous context manager built by the decorator. The value of data will be bound to the data variable after the as clause in the async with statement below.

❺ Lines after the yield will become the __aexit__ coroutine method. Here, another blocking call is delegated to the thread executor.

❻ Use web_page with async with.

This is very similar to the sequential @contextmanager decorator. Please see "Using @contextmanager" on page 668 for more details, including error handling at the yield line. For another example of @asynccontextmanager, see the contextlib documentation (*https://fpy.li/21-46*).

Now let's wrap up our coverage of asynchronous generator functions by contrasting them with native coroutines.

Asynchronous generators versus native coroutines

Here are some key similarities and differences between a native coroutine and an asynchronous generator function:

- Both are declared with async def.
- An asynchronous generator always has a yield expression in its body—that's what makes it a generator. A native coroutine never contains yield.
- A native coroutine may return some value other than None. An asynchronous generator can only use empty return statements.
- Native coroutines are awaitable: they can be driven by await expressions or passed to one of the many asyncio functions that take awaitable arguments, such as create_task. Asynchronous generators are not awaitable. They are asynchronous iterables, driven by async for or by asynchronous comprehensions.

Time to talk about asynchronous comprehensions.

Async Comprehensions and Async Generator Expressions

PEP 530—Asynchronous Comprehensions (*https://fpy.li/pep530*) introduced the use of async for and await in the syntax of comprehensions and generator expressions, starting with Python 3.6.

The only construct defined by PEP 530 that can appear outside an `async def` body is an asynchronous generator expression.

Defining and using an asynchronous generator expression

Given the `multi_probe` asynchronous generator from Example 21-18, we could write another asynchronous generator returning only the names of the domains found. Here is how—again using the asynchronous console launched with `-m asyncio`:

```
>>> from domainlib import multi_probe
>>> names = 'python.org rust-lang.org golang.org no-lang.invalid'.split()
>>> gen_found = (name async for name, found in multi_probe(names) if found)  ❶
>>> gen_found
<async_generator object <genexpr> at 0x10a8f9700>  ❷
>>> async for name in gen_found:  ❸
...     print(name)
...
golang.org
python.org
rust-lang.org
```

❶ The use of `async for` makes this an asynchronous generator expression. It can be defined anywhere in a Python module.

❷ The asynchronous generator expression builds an `async_generator` object— exactly the same type of object returned by an asynchronous generator function like `multi_probe`.

❸ The asynchronous generator object is driven by the `async for` statement, which in turn can only appear inside an `async def` body or in the magic asynchronous console I used in this example.

To summarize: an asynchronous generator expression can be defined anywhere in your program, but it can only be consumed inside a native coroutine or asynchronous generator function.

The remaining constructs introduced by PEP 530 can only be defined and used inside native coroutines or asynchronous generator functions.

Asynchronous comprehensions

Yury Selivanov—the author of PEP 530—justifies the need for asynchronous comprehensions with three short code snippets reproduced next.

We can all agree that we should be able to rewrite this code:

```
result = []
async for i in aiter():
```

```
    if i % 2:
        result.append(i)
```

like this:

```
result = [i async for i in aiter() if i % 2]
```

In addition, given a native coroutine fun, we should be able to write this:

```
result = [await fun() for fun in funcs]
```

 Using `await` in a list comprehension is similar to using `asyn cio.gather`. But `gather` gives you more control over exception handling, thanks to its optional `return_exceptions` argument. Caleb Hattingh recommends always setting `return_excep tions=True` (the default is `False`). Please see the `asyncio.gather` documentation (*https://fpy.li/21-48*) for more.

Back to the magic asynchronous console:

```
>>> names = 'python.org rust-lang.org golang.org no-lang.invalid'.split()
>>> names = sorted(names)
>>> coros = [probe(name) for name in names]
>>> await asyncio.gather(*coros)
[Result(domain='golang.org', found=True),
Result(domain='no-lang.invalid', found=False),
Result(domain='python.org', found=True),
Result(domain='rust-lang.org', found=True)]
>>> [await probe(name) for name in names]
[Result(domain='golang.org', found=True),
Result(domain='no-lang.invalid', found=False),
Result(domain='python.org', found=True),
Result(domain='rust-lang.org', found=True)]
>>>
```

Note that I sorted the list of names to show that the results come out in the order they were submitted, in both cases.

PEP 530 allows the use of `async for` and `await` in list comprehensions as well as in `dict` and `set` comprehensions. For example, here is a `dict` comprehension to store the results of `multi_probe` in the asynchronous console:

```
>>> {name: found async for name, found in multi_probe(names)}
{'golang.org': True, 'python.org': True, 'no-lang.invalid': False,
'rust-lang.org': True}
```

We can use the `await` keyword in the expression before the `for` or `async for` clause, and also in the expression after the `if` clause. Here is a set comprehension in the asynchronous console, collecting only the domains that were found:

```
>>> {name for name in names if (await probe(name)).found}
{'rust-lang.org', 'python.org', 'golang.org'}
```

I had to put extra parentheses around the `await` expression due to the higher precedence of the `__getattr__` operator . (dot).

Again, all of these comprehensions can only appear inside an `async def` body or in the enchanted asynchronous console.

Now let's talk about a very important feature of the `async` statements, `async` expressions, and the objects they create. Those constructs are often used with *asyncio* but, they are actually library independent.

async Beyond asyncio: Curio

Python's `async/await` language constructs are not tied to any specific event loop or library.[17] Thanks to the extensible API provided by special methods, anyone sufficiently motivated can write their own asynchronous runtime environment and framework to drive native coroutines, asynchronous generators, etc.

That's what David Beazley did in his *Curio* (*https://fpy.li/21-49*) project. He was interested in rethinking how these new language features could be used in a framework built from scratch. Recall that `asyncio` was released in Python 3.4, and it used `yield from` instead of `await`, so its API could not leverage asynchronous context managers, asynchronous iterators, and everything else that the `async/await` keywords made possible. As a result, *Curio* has a cleaner API and a simpler implementation, compared to `asyncio`.

Example 21-21 shows the *blogdom.py* script (Example 21-1) rewritten to use *Curio*.

Example 21-21. blogdom.py: Example 21-1, now using Curio

```
#!/usr/bin/env python3
from curio import run, TaskGroup
import curio.socket as socket
from keyword import kwlist

MAX_KEYWORD_LEN = 4


async def probe(domain: str) -> tuple[str, bool]:  ❶
    try:
        await socket.getaddrinfo(domain, None)  ❷
    except socket.gaierror:
        return (domain, False)
    return (domain, True)
```

---

17 That's in contrast with JavaScript, where `async/await` is hardwired to the built-in event loop and runtime environment, i.e., a browser, Node.js, or Deno.

```
async def main() -> None:
    names = (kw for kw in kwlist if len(kw) <= MAX_KEYWORD_LEN)
    domains = (f'{name}.dev'.lower() for name in names)
    async with TaskGroup() as group:    ❸
        for domain in domains:
            await group.spawn(probe, domain)    ❹
        async for task in group:    ❺
            domain, found = task.result
            mark = '+' if found else ' '
            print(f'{mark} {domain}')

if __name__ == '__main__':
    run(main())    ❻
```

❶ probe doesn't need to get the event loop, because…

❷ …getaddrinfo is a top-level function of curio.socket, not a method of a loop object—as it is in asyncio.

❸ A TaskGroup is a core concept in *Curio*, to monitor and control several coroutines, and to make sure they are all executed and cleaned up.

❹ TaskGroup.spawn is how you start a coroutine, managed by a specific TaskGroup instance. The coroutine is wrapped by a Task.

❺ Iterating with async for over a TaskGroup yields Task instances as each is completed. This corresponds to the line in Example 21-1 using for … as_completed(…):.

❻ *Curio* pioneered this sensible way to start an asynchronous program in Python.

To expand on the last point: if you look at the asyncio code examples for the first edition of *Fluent Python*, you'll see lines like these, repeated over and over:

```
loop = asyncio.get_event_loop()
loop.run_until_complete(main())
loop.close()
```

A *Curio* TaskGroup is an asynchronous context manager that replaces several ad hoc APIs and coding patterns in asyncio. We just saw how iterating over a TaskGroup makes the asyncio.as_completed(…) function unnecessary. Another example: instead of a special gather function, this snippet from the "Task Groups" docs (*https://fpy.li/21-50*) collects the results of all tasks in the group:

```
async with TaskGroup(wait=all) as g:
    await g.spawn(coro1)
    await g.spawn(coro2)
```

```
    await g.spawn(coro3)
  print('Results:', g.results)
```

Task groups support *structured concurrency* (*https://fpy.li/21-51*): a form of concurrent programming that constrains all the activity of a group of asynchronous tasks to a single entry and exit point. This is analogous to structured programming, which eschewed the GOTO command and introduced block statements to limit the entry and exit points of loops and subroutines. When used as an asynchronous context manager, a TaskGroup ensures that all tasks spawned inside are completed or cancelled, and any exceptions raised, upon exiting the enclosed block.

 Structured concurrency will probably be adopted by asyncio in upcoming Python releases. A strong indication appears in PEP 654–Exception Groups and except* (*https://fpy.li/pep654*), which was approved for Python 3.11 (*https://fpy.li/21-52*). The "Motivation" section (*https://fpy.li/21-53*) mentions *Trio's* "nurseries," their name for task groups: "Implementing a better task spawning API in *asyncio*, inspired by Trio nurseries, was the main motivation for this PEP."

Another important feature of *Curio* is better support for programming with coroutines and threads in the same codebase—a necessity in most nontrivial asynchronous programs. Starting a thread with await spawn_thread(func, …) returns an AsyncThread object with a Task-like interface. Threads can call coroutines thanks to a special AWAIT(coro) (*https://fpy.li/21-54*) function—named in all caps because await is now a keyword.

Curio also provides a UniversalQueue that can be used to coordinate the work among threads, *Curio* coroutines, and asyncio coroutines. That's right, *Curio* has features that allow it to run in a thread along with asyncio in another thread, in the same process, communicating via UniversalQueue and UniversalEvent. The API for these "universal" classes is the same inside and outside of coroutines, but in a coroutine, you need to prefix calls with await.

As I write this in October 2021, *HTTPX* is the first HTTP client library compatible with *Curio* (*https://fpy.li/21-55*), but I don't know of any asynchronous database libraries that support it yet. In the *Curio* repository there is an impressive set of network programming examples (*https://fpy.li/21-56*), including one using *WebSocket*, and another implementing the RFC 8305—Happy Eyeballs (*https://fpy.li/21-57*) concurrent algorithm for connecting to IPv6 endpoints with fast fallback to IPv4 if needed.

The design of *Curio* has been influential. The *Trio* (*https://fpy.li/21-58*) framework started by Nathaniel J. Smith was heavily inspired by *Curio*. *Curio* may also have prompted Python contributors to improve the usability of the asyncio API. For example, in its earliest releases, asyncio users very often had to get and pass around a loop object because some essential functions were either loop methods or required a loop argument. In recent versions of Python, direct access to the loop is not needed as often, and in fact several functions that accepted an optional loop are now deprecating that argument.

Type annotations for asynchronous types are our next topic.

Type Hinting Asynchronous Objects

The return type of a native coroutine describes what you get when you await on that coroutine, which is the type of the object that appears in the return statements in the body of the native coroutine function.[18]

This chapter provided many examples of annotated native coroutines, including probe from Example 21-21:

```python
async def probe(domain: str) -> tuple[str, bool]:
    try:
        await socket.getaddrinfo(domain, None)
    except socket.gaierror:
        return (domain, False)
    return (domain, True)
```

If you need to annotate a parameter that takes a coroutine object, then the generic type is:

```python
class typing.Coroutine(Awaitable[V_co], Generic[T_co, T_contra, V_co]):
    ...
```

That type, and the following types were introduced in Python 3.5 and 3.6 to annotate asynchronous objects:

```python
class typing.AsyncContextManager(Generic[T_co]):
    ...
class typing.AsyncIterable(Generic[T_co]):
    ...
class typing.AsyncIterator(AsyncIterable[T_co]):
    ...
class typing.AsyncGenerator(AsyncIterator[T_co], Generic[T_co, T_contra]):
    ...
class typing.Awaitable(Generic[T_co]):
    ...
```

With Python ≥ 3.9, use the collections.abc equivalents of these.

---

18 This differs from the annotations of classic coroutines, as discussed in "Generic Type Hints for Classic Coroutines" on page 654.

I want to highlight three aspects of those generic types.

First: they are all covariant on the first type parameter, which is the type of the items yielded from these objects. Recall rule #1 of "Variance rules of thumb" on page 555:

> If a formal type parameter defines a type for data that comes out of the object, it can be covariant.

Second: `AsyncGenerator` and `Coroutine` are contravariant on the second to last parameter. That's the type of the argument of the low-level `.send()` method that the event loop calls to drive asynchronous generators and coroutines. As such, it is an "input" type. Therefore, it can be contravariant, per Variance Rule of Thumb #2:

> If a formal type parameter defines a type for data that goes into the object after its initial construction, it can be contravariant.

Third: `AsyncGenerator` has no return type, in contrast with `typing.Generator`, which we saw in "Generic Type Hints for Classic Coroutines" on page 654. Returning a value by raising `StopIteration(value)` was one of the hacks that enabled generators to operate as coroutines and support `yield from`, as we saw in "Classic Coroutines" on page 645. There is no such overlap among the asynchronous objects: `AsyncGenerator` objects don't return values, and are completely separate from native coroutine objects, which are annotated with `typing.Coroutine`.

Finally, let's briefly discuss the advantages and challenges of asynchronous programming.

How Async Works and How It Doesn't

The sections closing this chapter discuss high-level ideas around asynchronous programming, regardless of the language or library you are using.

Let's begin by explaining the #1 reason why asynchronous programming is appealing, followed by a popular myth, and how to deal with it.

Running Circles Around Blocking Calls

Ryan Dahl, the inventor of Node.js, introduces the philosophy of his project by saying "We're doing I/O completely wrong."[19] He defines a *blocking function* as one that does file or network I/O, and argues that we can't treat them as we treat nonblocking functions. To explain why, he presents the numbers in the second column of Table 21-1.

---

19 Video: "Introduction to Node.js" (*https://fpy.li/21-59*) at 4:55.

Table 21-1. Modern computer latency for reading data from different devices; third column shows proportional times in a scale easier to understand for us slow humans

Device	CPU cycles	Proportional "human" scale
L1 cache	3	3 seconds
L2 cache	14	14 seconds
RAM	250	250 seconds
disk	41,000,000	1.3 years
network	240,000,000	7.6 years

To make sense of Table 21-1, bear in mind that modern CPUs with GHz clocks run billions of cycles per second. Let's say that a CPU runs exactly 1 billion cycles per second. That CPU can make more than 333 million L1 cache reads in 1 second, or 4 (four!) network reads in the same time. The third column of Table 21-1 puts those numbers in perspective by multiplying the second column by a constant factor. So, in an alternate universe, if one read from L1 cache took 3 seconds, then a network read would take 7.6 years!

Table 21-1 explains why a disciplined approach to asynchronous programming can lead to high-performance servers. The challenge is achieving that discipline. The first step is to recognize that "I/O bound system" is a fantasy.

The Myth of I/O-Bound Systems

A commonly repeated meme is that asynchronous programming is good for "I/O bound systems." I learned the hard way that there are no "I/O-bound systems." You may have I/O-bound *functions*. Perhaps the vast majority of the functions in your system are I/O bound; i.e., they spend more time waiting for I/O than crunching data. While waiting, they cede control to the event loop, which can then drive some other pending task. But inevitably, any nontrivial system will have some parts that are CPU bound. Even trivial systems reveal that, under stress. In "Soapbox" on page 834, I tell the story of two asynchronous programs that struggled with CPU-bound functions slowing down the event loop with severe impact on performance.

Given that any nontrivial system will have CPU-bound functions, dealing with them is the key to success in asynchronous programming.

Avoiding CPU-Bound Traps

If you're using Python at scale, you should have some automated tests designed specifically to detect performance regressions as soon as they appear. This is critically important with asynchronous code, but also relevant to threaded Python code— because of the GIL. If you wait until the slowdown starts bothering the development team, it's too late. The fix will probably require some major makeover.

Here are some options for when you identify a CPU-hogging bottleneck:

- Delegate the task to a Python process pool.
- Delegate the task to an external task queue.
- Rewrite the relevant code in Cython, C, Rust, or some other language that compiles to machine code and interfaces with the Python/C API, preferably releasing the GIL.
- Decide that you can afford the performance hit and do nothing—but record the decision to make it easier to revert to it later.

The external task queue should be chosen and integrated as soon as possible at the start of the project, so that nobody in the team hesitates to use it when needed.

The last option—do nothing—falls in the category of technical debt (*https://fpy.li/ 21-60*).

Concurrent programming is a fascinating topic, and I would like to write a lot more about it. But it is not the main focus of this book, and this is already one of the longest chapters, so let's wrap it up.

Chapter Summary

> The problem with normal approaches to asynchronous programming is that they're all-or-nothing propositions. You rewrite all your code so none of it blocks or you're just wasting your time.
>
> —Alvaro Videla and Jason J. W. Williams, *RabbitMQ in Action*

I chose that epigraph for this chapter for two reasons. At a high level, it reminds us to avoid blocking the event loop by delegating slow tasks to a different processing unit, from a simple thread all the way to a distributed task queue. At a lower level, it is also a warning: once you write your first `async def`, your program is inevitably going to have more and more `async def`, `await`, `async with`, and `async for`. And using non-asynchronous libraries suddenly becomes a challenge.

After the simple *spinner* examples in Chapter 19, here our main focus was asynchronous programming with native coroutines, starting with the *blogdom.py* DNS probing example, followed by the concept of *awaitables*. While reading the source code of *flags_asyncio.py*, we found the first example of an *asynchronous context manager*.

The more advanced variations of the flag downloading program introduced two powerful functions: the `asyncio.as_completed` generator and the `loop.run_in_exec utor` coroutine. We also saw the concept and application of a semaphore to limit the number of concurrent downloads—as expected from well-behaved HTTP clients.

Server-side asynchronous programming was presented through the *mojifinder* examples: a *FastAPI* web service and *tcp_mojifinder.py*—the latter using just asyncio and the TCP protocol.

Asynchronous iteration and asynchronous iterables were the next major topic, with sections on async for, Python's async console, asynchronous generators, asynchronous generator expressions, and asynchronous comprehensions.

The last example in the chapter was *blogdom.py* rewritten with the *Curio* framework, to demonstrate how Python's asynchronous features are not tied to the asyncio package. *Curio* also showcases the concept of *structured concurrency*, which may have an industry-wide impact, bringing more clarity to concurrent code.

Finally, the sections under "How Async Works and How It Doesn't" on page 829 discuss the main appeal of asynchronous programming, the misconception of "I/O-bound systems," and dealing with the inevitable CPU-bound parts of your program.

Further Reading

David Beazley's PyOhio 2016 keynote "Fear and Awaiting in Async" (*https://fpy.li/ 21-61*) is a fantastic, live-coded introduction to the potential of the language features made possible by Yury Selivanov's contribution of the async/await keywords in Python 3.5. At one point, Beazley complains that await can't be used in list comprehensions, but that was fixed by Selivanov in PEP 530—Asynchronous Comprehensions (*https://fpy.li/pep530*), implemented in Python 3.6 later in that same year. Apart from that, everything else in Beazley's keynote is timeless, as he demonstrates how the asynchronous objects we saw in this chapter work, without the help of any framework—just a simple run function using .send(None) to drive coroutines. Only at the very end Beazley shows *Curio* (*https://fpy.li/21-62*), which he started that year as an experiment to see how far can you go doing asynchronous programming without a foundation of callbacks or futures, just coroutines. As it turns out, you can go very far —as demonstrated by the evolution of *Curio* and the later creation of *Trio* (*https:// fpy.li/21-58*) by Nathaniel J. Smith. *Curio's* documentation has links (*https://fpy.li/ 21-64*) to more talks by Beazley on the subject.

Besides starting *Trio*, Nathaniel J. Smith wrote two deep blog posts that I highly recommend: "Some thoughts on asynchronous API design in a post-async/await world" (*https://fpy.li/21-65*), contrasting the design of *Curio* with that of *asyncio*,and "Notes on structured concurrency, or: Go statement considered harmful" (*https://fpy.li/ 21-66*), about structured concurrency. Smith also gave a long and informative answer to the question: "What is the core difference between asyncio and trio?" (*https:// fpy.li/21-67*) on StackOverflow.

To learn more about the *asyncio* package, I've mentioned the best written resources I know at the start of this chapter: the official documentation (*https://fpy.li/21-1*) after the outstanding overhaul (*https://fpy.li/21-69*) started by Yury Selivanov in 2018, and Caleb Hattingh's book *Using Asyncio in Python* (O'Reilly). In the official documentation, make sure to read "Developing with asyncio" (*https://fpy.li/21-70*): documenting the *asyncio* debug mode, and also discussing common mistakes and traps and how to avoid them.

For a very accessible, 30-minute introduction to asynchronous programming in general and also *asyncio*, watch Miguel Grinberg's "Asynchronous Python for the Complete Beginner" (*https://fpy.li/21-71*), presented at PyCon 2017. Another great introduction is "Demystifying Python's Async and Await Keywords" (*https://fpy.li/21-72*), presented by Michael Kennedy, where among other things I learned about the *unsync* (*https://fpy.li/21-73*) library that provides a decorator to delegate the execution of coroutines, I/O-bound functions, and CPU-bound functions to asyncio, threading, or multiprocessing as needed.

At EuroPython 2019, Lynn Root—a global leader of *PyLadies* (*https://fpy.li/21-74*)—presented the excellent "Advanced asyncio: Solving Real-world Production Problems" (*https://fpy.li/21-75*), informed by her experience using Python as a staff engineer at Spotify.

In 2020, Łukasz Langa recorded a series of great videos about *asyncio*, starting with "Learn Python's AsyncIO #1—The Async Ecosystem" (*https://fpy.li/21-76*). Langa also made the super cool video "AsyncIO + Music" (*https://fpy.li/21-77*) for PyCon 2020 that not only shows *asyncio* applied in a very concrete event-oriented domain, but also explains it from the ground up.

Another area dominated by event-oriented programming is embedded systems. That's why Damien George added support for async/await in his *MicroPython* (*https://fpy.li/21-78*) interpreter for microcontrollers. At PyCon Australia 2018, Matt Trentini demonstrated the *uasyncio* (*https://fpy.li/21-79*) library, a subset of *asyncio* that is part of MicroPython's standard library.

For higher-level thinking about async programming in Python, read the blog post "Python async frameworks—Beyond developer tribalism" (*https://fpy.li/21-80*) by Tom Christie.

Finally, I recommend "What Color Is Your Function?" (*https://fpy.li/21-81*) by Bob Nystrom, discussing the incompatible execution models of plain functions versus async functions—a.k.a. coroutines—in JavaScript, Python, C#, and other languages. Spoiler alert: Nystrom's conclusion is that the language that got this right is Go, where all functions are the same color. I like that about Go. But I also think Nathaniel J. Smith has a point when he wrote "Go statement considered harmful" (*https://fpy.li/21-66*). Nothing is perfect, and concurrent programming is always hard.

Soapbox

How a Slow Function Almost Spoiled the uvloop Benchmarks

In 2016, Yury Selivanov released *uvloop* (*https://fpy.li/21-83*), "a fast, drop-in replacement of the built-in *asyncio* event loop." The benchmarks presented in Selivanov's blog post (*https://fpy.li/21-84*) announcing the library in 2016 are very impressive. He wrote: "it is at least 2x faster than nodejs, gevent, as well as any other Python asynchronous framework. The performance of uvloop-based asyncio is close to that of Go programs."

However, the post reveals that *uvloop* is able to match the performance of Go under two conditions:

1. Go is configured to use a single thread. That makes the Go runtime behave similarly to *asyncio*: concurrency is achieved via multiple coroutines driven by an event loop, all in the same thread.[20]

2. The Python 3.5 code uses *httptools* (*https://fpy.li/21-85*) in addition to *uvloop* itself.

Selivanov explains that he wrote *httptools* after benchmarking *uvloop* with *aiohttp* (*https://fpy.li/21-86*)—one of the first full-featured HTTP libraries built on `asyncio`:

> However, the performance bottleneck in *aiohttp* turned out to be its HTTP parser, which is so slow, that it matters very little how fast the underlying I/O library is. To make things more interesting, we created a Python binding for *http-parser* (Node.js HTTP parser C library, originally developed for *NGINX*). The library is called *httptools*, and is available on Github and PyPI.

Now think about that: Selivanov's HTTP performance tests consisted of a simple echo server written in the different languages/libraries, pounded by the *wrk* (*https://fpy.li/21-87*) benchmarking tool. Most developers would consider a simple echo server an "I/O-bound system," right? But it turned out that parsing HTTP headers is CPU bound, and it had a slow Python implementation in *aiohttp* in when Selivanov did the benchmarks in 2016. Whenever a function written in Python was parsing headers, the event loop was blocked. The impact was so significant that Selivanov went to the extra trouble of writing *httptools*. Without optimizing the CPU-bound code, the performance gains of a faster event loop were lost.

---

20 Using a single thread was the default setting until Go 1.5 was released. Years before, Go had already earned a well-deserved reputation for enabling highly concurrent networked systems. One more evidence that concurrency doesn't require multiple threads or CPU cores.

Death by a Thousand Cuts

Instead of a simple echo server, imagine a complex and evolving Python system with tens of thousands of lines of asynchronous code, interfacing with many external libraries. Years ago I was asked to help diagnose performance problems in a system like that. It was written in Python 2.7 with the *Twisted* (*https://fpy.li/21-88*) framework—a solid library and in many ways a precursor to `asyncio` itself.

Python was used to build a façade for the web UI, integrating functionality provided by preexisting libraries and command-line tools written in other languages—but not designed for concurrent execution.

The project was ambitious; it had been in development for more than a year already, but it was not in production yet.[21] Over time, the developers noticed that the performance of the whole system was decreasing, and they were having a hard time finding the bottlenecks.

What was happening: with each added feature, more CPU-bound code was slowing down *Twisted*'s event loop. Python's role as a glue language meant there was a lot of data parsing and conversion between formats. There wasn't a single bottleneck: the problem was spread over countless little functions added over months of development. Fixing that would require rethinking the architecture of the system, rewriting a lot of code, probably leveraging a task queue, and perhaps using microservices or custom libraries written in languages better suited for CPU-intensive concurrent processing. The stakeholders were not prepared to make that additional investment, and the project was cancelled shortly afterwards.

When I told this story to Glyph Lefkowitz—founder the *Twisted* project—he said that one of his priorities at the start of an asynchronous programming project is to decide which tools he will use to farm out the CPU-intensive tasks. This conversation with Glyph was the inspiration for "Avoiding CPU-Bound Traps" on page 830.

---

21 Regardless of technical choices, this was probably the biggest mistake in this project: the stakeholders did not go for an MVP approach—delivering a Minimum Viable Product as soon as possible, and then adding features at a steady pace.

## PART V

Metaprogramming

## CHAPTER 22

Dynamic Attributes and Properties

> The crucial importance of properties is that their existence makes it perfectly safe and indeed advisable for you to expose public data attributes as part of your class's public interface.
>
> — Martelli, Ravenscroft, and Holden, "Why properties are important"[1]

Data attributes and methods are collectively known as *attributes* in Python. A method is an attribute that is *callable*. *Dynamic attributes* present the same interface as data attributes—i.e., `obj.attr`—but are computed on demand. This follows Bertrand Meyer's *Uniform Access Principle*:

> All services offered by a module should be available through a uniform notation, which does not betray whether they are implemented through storage or through computation.[2]

There are several ways to implement dynamic attributes in Python. This chapter covers the simplest ways: the `@property` decorator and the `__getattr__` special method.

A user-defined class implementing `__getattr__` can implement a variation of dynamic attributes that I call *virtual attributes*: attributes that are not explicitly declared anywhere in the source code of the class, and are not present in the instance `__dict__`, but may be retrieved elsewhere or computed on the fly whenever a user tries to read a nonexistent attribute like `obj.no_such_attr`.

Coding dynamic and virtual attributes is the kind of metaprogramming that framework authors do. However, in Python the basic techniques are straightforward, so we can use them in everyday data wrangling tasks. That's how we'll start this chapter.

---

1 Alex Martelli, Anna Ravenscroft, and Steve Holden, *Python in a Nutshell*, 3rd ed. (O'Reilly), p. 123.

2 Bertrand Meyer, *Object-Oriented Software Construction*, 2nd ed. (Pearson), p. 57.

What's New in This Chapter

Most of the updates to this chapter were motivated by a discussion of `@func tools.cached_property` (introduced in Python 3.8), as well as the combined use of `@property` with `@functools.cache` (new in 3.9). This affected the code for the Record and Event classes that appear in "Computed Properties" on page 849. I also added a refactoring to leverage the PEP 412—Key-Sharing Dictionary (*https://fpy.li/ pep412*) optimization.

To highlight more relevant features while keeping the examples readable, I removed some nonessential code—merging the old DbRecord class into Record, replacing `shelve.Shelve` with a dict, and deleting the logic to download the OSCON dataset —which the examples now read from a local file included in the *Fluent Python* code repository (*https://fpy.li/code*).

Data Wrangling with Dynamic Attributes

In the next few examples, we'll leverage dynamic attributes to work with a JSON dataset published by O'Reilly for the OSCON 2014 conference. Example 22-1 shows four records from that dataset.[3]

Example 22-1. Sample records from osconfeed.json; some field contents abbreviated

```
{ "Schedule":
  { "conferences": [{"serial": 115 }],
    "events": [
      { "serial": 34505,
        "name": "Why Schools Don't Use Open Source to Teach Programming",
        "event_type": "40-minute conference session",
        "time_start": "2014-07-23 11:30:00",
        "time_stop": "2014-07-23 12:10:00",
        "venue_serial": 1462,
        "description": "Aside from the fact that high school programming...",
        "website_url": "http://oscon.com/oscon2014/public/schedule/detail/34505",
        "speakers": [157509],
        "categories": ["Education"] }
    ],
    "speakers": [
      { "serial": 157509,
        "name": "Robert Lefkowitz",
        "photo": null,
        "url": "http://sharewave.com/",
```

---

3 OSCON—O'Reilly Open Source Conference—was a casualty of the COVID-19 pandemic. The original 744 KB JSON file I used for these examples is no longer online as of January 10, 2021. You'll find a copy of *oscon-feed.json* in the example code repository (*https://fpy.li/22-1*).

```
            "position": "CTO",
            "affiliation": "Sharewave",
            "twitter": "sharewaveteam",
            "bio": "Robert ´r0ml´ Lefkowitz is the CTO at Sharewave, a startup..." }
    ],
    "venues": [
      { "serial": 1462,
        "name": "F151",
        "category": "Conference Venues" }
    ]
  }
}
```

Example 22-1 shows 4 of the 895 records in the JSON file. The entire dataset is a single JSON object with the key "Schedule", and its value is another mapping with four keys: "conferences", "events", "speakers", and "venues". Each of those four keys maps to a list of records. In the full dataset, the "events", "speakers", and "venues" lists have dozens or hundreds of records, while "conferences" has only that one record shown in Example 22-1. Every record has a "serial" field, which is a unique identifier for the record within the list.

I used Python's console to explore the dataset, as shown in Example 22-2.

Example 22-2. Interactive exploration of osconfeed.json

```
>>> import json
>>> with open('data/osconfeed.json') as fp:
...     feed = json.load(fp)          ❶
>>> sorted(feed['Schedule'].keys())   ❷
['conferences', 'events', 'speakers', 'venues']
>>> for key, value in sorted(feed['Schedule'].items()):
...     print(f'{len(value):3} {key}')  ❸
...
  1 conferences
484 events
357 speakers
 53 venues
>>> feed['Schedule']['speakers'][-1]['name']   ❹
'Carina C. Zona'
>>> feed['Schedule']['speakers'][-1]['serial']  ❺
141590
>>> feed['Schedule']['events'][40]['name']
'There *Will* Be Bugs'
>>> feed['Schedule']['events'][40]['speakers']  ❻
[3471, 5199]
```

❶   feed is a dict holding nested dicts and lists, with string and integer values.

❷   List the four record collections inside "Schedule".

❸ Display record counts for each collection.

❹ Navigate through the nested dicts and lists to get the name of the last speaker.

❺ Get the serial number of that same speaker.

❻ Each event has a `'speakers'` list with zero or more speaker serial numbers.

Exploring JSON-Like Data with Dynamic Attributes

Example 22-2 is simple enough, but the syntax `feed['Schedule']['events'][40]`
`['name']` is cumbersome. In JavaScript, you can get the same value by writing
`feed.Schedule.events[40].name`. It's easy to implement a `dict`-like class that does
the same in Python—there are plenty of implementations on the web.[4] I wrote
`FrozenJSON`, which is simpler than most recipes because it supports reading only: it's
just for exploring the data. `FrozenJSON` is also recursive, dealing automatically with
nested mappings and lists.

Example 22-3 is a demonstration of `FrozenJSON`, and the source code is shown in
Example 22-4.

*Example 22-3. FrozenJSON from Example 22-4 allows reading attributes like name, and
calling methods like .keys() and .items()*

```
>>> import json
>>> raw_feed = json.load(open('data/osconfeed.json'))
>>> feed = FrozenJSON(raw_feed)  ❶
>>> len(feed.Schedule.speakers)  ❷
357
>>> feed.keys()
dict_keys(['Schedule'])
>>> sorted(feed.Schedule.keys())  ❸
['conferences', 'events', 'speakers', 'venues']
>>> for key, value in sorted(feed.Schedule.items()):  ❹
...     print(f'{len(value):3} {key}')
...
  1 conferences
484 events
357 speakers
 53 venues
>>> feed.Schedule.speakers[-1].name  ❺
'Carina C. Zona'
>>> talk = feed.Schedule.events[40]
>>> type(talk)  ❻
```

---

4 Two examples are `AttrDict` (*https://fpy.li/22-2*) and `addict` (*https://fpy.li/22-3*).

```
<class 'explore0.FrozenJSON'>
>>> talk.name
'There *Will* Be Bugs'
>>> talk.speakers  ❼
[3471, 5199]
>>> talk.flavor  ❽
Traceback (most recent call last):
  ...
KeyError: 'flavor'
```

❶ Build a FrozenJSON instance from the raw_feed made of nested dicts and lists.

❷ FrozenJSON allows traversing nested dicts by using attribute notation; here we show the length of the list of speakers.

❸ Methods of the underlying dicts can also be accessed, like .keys(), to retrieve the record collection names.

❹ Using items(), we can retrieve the record collection names and their contents, to display the len() of each of them.

❺ A list, such as feed.Schedule.speakers, remains a list, but the items inside are converted to FrozenJSON if they are mappings.

❻ Item 40 in the events list was a JSON object; now it's a FrozenJSON instance.

❼ Event records have a speakers list with speaker serial numbers.

❽ Trying to read a missing attribute raises KeyError, instead of the usual AttributeError.

The keystone of the FrozenJSON class is the __getattr__ method, which we already used in the Vector example in "Vector Take #3: Dynamic Attribute Access" on page 409, to retrieve Vector components by letter: v.x, v.y, v.z, etc. It's essential to recall that the __getattr__ special method is only invoked by the interpreter when the usual process fails to retrieve an attribute (i.e., when the named attribute cannot be found in the instance, nor in the class or in its superclasses).

The last line of Example 22-3 exposes a minor issue with my code: trying to read a missing attribute should raise AttributeError, and not KeyError as shown. When I implemented the error handling to do that, the __getattr__ method became twice as long, distracting from the most important logic I wanted to show. Given that users would know that a FrozenJSON is built from mappings and lists, I think the KeyError is not too confusing.

Example 22-4. explore0.py: turn a JSON dataset into a FrozenJSON holding nested FrozenJSON objects, lists, and simple types

```
from collections import abc


class FrozenJSON:
    """A read-only façade for navigating a JSON-like object
       using attribute notation
    """

    def __init__(self, mapping):
        self.__data = dict(mapping)  ❶

    def __getattr__(self, name):  ❷
        try:
            return getattr(self.__data, name)  ❸
        except AttributeError:
            return FrozenJSON.build(self.__data[name])  ❹

    def __dir__(self):  ❺
        return self.__data.keys()

    @classmethod
    def build(cls, obj):  ❻
        if isinstance(obj, abc.Mapping):  ❼
            return cls(obj)
        elif isinstance(obj, abc.MutableSequence):  ❽
            return [cls.build(item) for item in obj]
        else:  ❾
            return obj
```

❶ Build a dict from the mapping argument. This ensures we get a mapping or something that can be converted to one. The double-underscore prefix on __data makes it a *private attribute*.

❷ __getattr__ is called only when there's no attribute with that name.

❸ If name matches an attribute of the instance __data dict, return that. This is how calls like feed.keys() are handled: the keys method is an attribute of the __data dict.

❹ Otherwise, fetch the item with the key `name` from `self.__data`, and return the result of calling `FrozenJSON.build()` on that.[5]

❺ Implementing `__dir__` suports the `dir()` built-in, which in turns supports auto-completion in the standard Python console as well as IPython, Jupyter Notebook, etc. This simple code will enable recursive auto-completion based on the keys in `self.__data`, because `__getattr__` builds `FrozenJSON` instances on the fly—useful for interactive exploration of the data.

❻ This is an alternate constructor, a common use for the `@classmethod` decorator.

❼ If `obj` is a mapping, build a `FrozenJSON` with it. This is an example of *goose typing*—see "Goose Typing" on page 444 if you need a refresher.

❽ If it is a `MutableSequence`, it must be a list,[6] so we build a `list` by passing each item in `obj` recursively to `.build()`.

❾ If it's not a `dict` or a `list`, return the item as it is.

A `FrozenJSON` instance has the `__data` private instance attribute stored under the name `_FrozenJSON__data`, as explained in "Private and 'Protected' Attributes in Python" on page 384. Attempts to retrieve attributes by other names will trigger `__getattr__`. This method will first look if the `self.__data` dict has an attribute (not a key!) by that name; this allows `FrozenJSON` instances to handle `dict` methods such as `items`, by delegating to `self.__data.items()`. If `self.__data` doesn't have an attribute with the given `name`, `__getattr__` uses `name` as a key to retrieve an item from `self.__data`, and passes that item to `FrozenJSON.build`. This allows navigating through nested structures in the JSON data, as each nested mapping is converted to another `FrozenJSON` instance by the `build` class method.

Note that `FrozenJSON` does not transform or cache the original dataset. As we traverse the data, `__getattr__` creates `FrozenJSON` instances again and again. That's OK for a dataset of this size, and for a script that will only be used to explore or convert the data.

---

5 The expression `self.__data[name]` is where a `KeyError` exception may occur. Ideally, it should be handled and an `AttributeError` raised instead, because that's what is expected from `__getattr__`. The diligent reader is invited to code the error handling as an exercise.

6 The source of the data is JSON, and the only collection types in JSON data are `dict` and `list`.

Any script that generates or emulates dynamic attribute names from arbitrary sources must deal with one issue: the keys in the original data may not be suitable attribute names. The next section addresses this.

The Invalid Attribute Name Problem

The FrozenJSON code doesn't handle attribute names that are Python keywords. For example, if you build an object like this:

```
>>> student = FrozenJSON({'name': 'Jim Bo', 'class': 1982})
```

You won't be able to read student.class because class is a reserved keyword in Python:

```
>>> student.class
  File "<stdin>", line 1
    student.class
                ^
SyntaxError: invalid syntax
```

You can always do this, of course:

```
>>> getattr(student, 'class')
1982
```

But the idea of FrozenJSON is to provide convenient access to the data, so a better solution is checking whether a key in the mapping given to FrozenJSON.__init__ is a keyword, and if so, append an _ to it, so the attribute can be read like this:

```
>>> student.class_
1982
```

This can be achieved by replacing the one-liner __init__ from Example 22-4 with the version in Example 22-5.

Example 22-5. explore1.py: append an _ to attribute names that are Python keywords

```
def __init__(self, mapping):
    self.__data = {}
    for key, value in mapping.items():
        if keyword.iskeyword(key):   ❶
            key += '_'
        self.__data[key] = value
```

❶ The keyword.iskeyword(...) function is exactly what we need; to use it, the key word module must be imported, which is not shown in this snippet.

A similar problem may arise if a key in a JSON record is not a valid Python identifier:

```
>>> x = FrozenJSON({'2be':'or not'})
>>> x.2be
```

```
File "<stdin>", line 1
    x.2be
      ^
SyntaxError: invalid syntax
```

Such problematic keys are easy to detect in Python 3 because the str class provides the s.isidentifier() method, which tells you whether s is a valid Python identifier according to the language grammar. But turning a key that is not a valid identifier into a valid attribute name is not trivial. One solution would be to implement __geti tem__ to allow attribute access using notation like x['2be']. For the sake of simplicity, I will not worry about this issue.

After giving some thought to the dynamic attribute names, let's turn to another essential feature of FrozenJSON: the logic of the build class method. Fro zen.JSON.build is used by __getattr__ to return a different type of object depending on the value of the attribute being accessed: nested structures are converted to FrozenJSON instances or lists of FrozenJSON instances.

Instead of a class method, the same logic could be implemented as the __new__ special method, as we'll see next.

Flexible Object Creation with __new__

We often refer to __init__ as the constructor method, but that's because we adopted jargon from other languages. In Python, __init__ gets self as the first argument, therefore the object already exists when __init__ is called by the interpreter. Also, __init__ cannot return anything. So it's really an initializer, not a constructor.

When a class is called to create an instance, the special method that Python calls on that class to construct an instance is __new__. It's a class method, but gets special treatment, so the @classmethod decorator is not applied to it. Python takes the instance returned by __new__ and then passes it as the first argument self of __init__. We rarely need to code __new__, because the implementation inherited from object suffices for the vast majority of use cases.

If necessary, the __new__ method can also return an instance of a different class. When that happens, the interpreter does not call __init__. In other words, Python's logic for building an object is similar to this pseudocode:

```
# pseudocode for object construction
def make(the_class, some_arg):
    new_object = the_class.__new__(some_arg)
    if isinstance(new_object, the_class):
        the_class.__init__(new_object, some_arg)
    return new_object

# the following statements are roughly equivalent
```

```
x = Foo('bar')
x = make(Foo, 'bar')
```

Example 22-6 shows a variation of FrozenJSON where the logic of the former build class method was moved to __new__.

Example 22-6. explore2.py: using __new__ instead of build to construct new objects that may or may not be instances of FrozenJSON

```
from collections import abc
import keyword

class FrozenJSON:
    """A read-only façade for navigating a JSON-like object
       using attribute notation
    """

    def __new__(cls, arg):  ❶
        if isinstance(arg, abc.Mapping):
            return super().__new__(cls)  ❷
        elif isinstance(arg, abc.MutableSequence):  ❸
            return [cls(item) for item in arg]
        else:
            return arg

    def __init__(self, mapping):
        self.__data = {}
        for key, value in mapping.items():
            if keyword.iskeyword(key):
                key += '_'
            self.__data[key] = value

    def __getattr__(self, name):
        try:
            return getattr(self.__data, name)
        except AttributeError:
            return FrozenJSON(self.__data[name])  ❹

    def __dir__(self):
        return self.__data.keys()
```

❶ As a class method, the first argument __new__ gets is the class itself, and the remaining arguments are the same that __init__ gets, except for self.

❷ The default behavior is to delegate to the __new__ of a superclass. In this case, we are calling __new__ from the object base class, passing FrozenJSON as the only argument.

❸ The remaining lines of __new__ are exactly as in the old build method.

❹ This was where `FrozenJSON.build` was called before; now we just call the `FrozenJSON` class, which Python handles by calling `FrozenJSON.__new__`.

The `__new__` method gets the class as the first argument because, usually, the created object will be an instance of that class. So, in `FrozenJSON.__new__`, when the expression `super().__new__(cls)` effectively calls `object.__new__(FrozenJSON)`, the instance built by the `object` class is actually an instance of `FrozenJSON`. The `__class__` attribute of the new instance will hold a reference to `FrozenJSON`, even though the actual construction is performed by `object.__new__`, implemented in C, in the guts of the interpreter.

The OSCON JSON dataset is structured in a way that is not helpful for interactive exploration. For example, the event at index 40, titled `'There *Will* Be Bugs'` has two speakers, 3471 and 5199. Finding the names of the speakers is awkward, because those are serial numbers and the `Schedule.speakers` list is not indexed by them. To get each speaker, we must iterate over that list until we find a record with a matching serial number. Our next task is restructuring the data to prepare for automatic retrieval of linked records.

Computed Properties

We first saw the `@property` decorator in Chapter 11, in the section, "A Hashable Vector2d" on page 376. In Example 11-7, I used two properties in `Vector2d` just to make the x and y attributes read-only. Here we will see properties that compute values, leading to a discussion of how to cache such values.

The records in the `'events'` list of the OSCON JSON data contain integer serial numbers pointing to records in the `'speakers'` and `'venues'` lists. For example, this is the record for a conference talk (with an elided description):

```
{ "serial": 33950,
  "name": "There *Will* Be Bugs",
  "event_type": "40-minute conference session",
  "time_start": "2014-07-23 14:30:00",
  "time_stop": "2014-07-23 15:10:00",
  "venue_serial": 1449,
  "description": "If you're pushing the envelope of programming...",
  "website_url": "http://oscon.com/oscon2014/public/schedule/detail/33950",
  "speakers": [3471, 5199],
  "categories": ["Python"] }
```

We will implement an `Event` class with `venue` and `speakers` properties to return the linked data automatically—in other words, "dereferencing" the serial number. Given an `Event` instance, Example 22-7 shows the desired behavior.

Example 22-7. Reading venue and speakers returns Record objects

```
>>> event  ❶
<Event 'There *Will* Be Bugs'>
>>> event.venue  ❷
<Record serial=1449>
>>> event.venue.name  ❸
'Portland 251'
>>> for spkr in event.speakers:  ❹
...     print(f'{spkr.serial}: {spkr.name}')
...
3471: Anna Martelli Ravenscroft
5199: Alex Martelli
```

❶ Given an Event instance…

❷ …reading event.venue returns a Record object instead of a serial number.

❸ Now it's easy to get the name of the venue.

❹ The event.speakers property returns a list of Record instances.

As usual, we will build the code step-by-step, starting with the Record class and a function to read the JSON data and return a dict with Record instances.

Step 1: Data-Driven Attribute Creation

Example 22-8 shows the doctest to guide this first step.

Example 22-8. Test-driving schedule_v1.py (from Example 22-9)

```
>>> records = load(JSON_PATH)  ❶
>>> speaker = records['speaker.3471']  ❷
>>> speaker  ❸
<Record serial=3471>
>>> speaker.name, speaker.twitter  ❹
('Anna Martelli Ravenscroft', 'annaraven')
```

❶ load a dict with the JSON data.

❷ The keys in records are strings built from the record type and serial number.

❸ speaker is an instance of the Record class defined in Example 22-9.

❹ Fields from the original JSON can be retrieved as Record instance attributes.

The code for *schedule_v1.py* is in Example 22-9.

Example 22-9. schedule_v1.py: reorganizing the OSCON schedule data

```python
import json

JSON_PATH = 'data/osconfeed.json'

class Record:
    def __init__(self, **kwargs):
        self.__dict__.update(kwargs)  ❶

    def __repr__(self):
        return f'<{self.__class__.__name__} serial={self.serial!r}>'  ❷

def load(path=JSON_PATH):
    records = {}  ❸
    with open(path) as fp:
        raw_data = json.load(fp)  ❹
    for collection, raw_records in raw_data['Schedule'].items():  ❺
        record_type = collection[:-1]  ❻
        for raw_record in raw_records:
            key = f'{record_type}.{raw_record["serial"]}'  ❼
            records[key] = Record(**raw_record)  ❽
    return records
```

❶ This is a common shortcut to build an instance with attributes created from keyword arguments (detailed explanation follows).

❷ Use the `serial` field to build the custom `Record` representation shown in Example 22-8.

❸ `load` will ultimately return a `dict` of `Record` instances.

❹ Parse the JSON, returning native Python objects: lists, dicts, strings, numbers, etc.

❺ Iterate over the four top-level lists named `'conferences'`, `'events'`, `'speakers'`, and `'venues'`.

❻ `record_type` is the list name without the last character, so `speakers` becomes `speaker`. In Python ≥ 3.9 we can do this more explicitly with `collection.removesuffix('s')`—see PEP 616—String methods to remove prefixes and suffixes (*https://fpy.li/pep616*).

❼ Build the key in the format `'speaker.3471'`.

❽ Create a `Record` instance and save it in `records` with the key.

The `Record.__init__` method illustrates an old Python hack. Recall that the `__dict__` of an object is where its attributes are kept—unless `__slots__` is declared in the class, as we saw in "Saving Memory with __slots__" on page 386. So, updating an instance `__dict__` with a mapping is a quick way to create a bunch of attributes in that instance.[7]

Depending on the application, the `Record` class may need to deal with keys that are not valid attribute names, as we saw in "The Invalid Attribute Name Problem" on page 846. Dealing with that issue would distract from the key idea of this example, and is not a problem in the dataset we are reading.

The definition of `Record` in Example 22-9 is so simple that you may be wondering why I did not use it before, instead of the more complicated `FrozenJSON`. There are two reasons. First, `FrozenJSON` works by recursively converting the nested mappings and lists; `Record` doesn't need that because our converted dataset doesn't have mappings nested in mappings or lists. The records contain only strings, integers, lists of strings, and lists of integers. Second reason: `FrozenJSON` provides access to the embedded `__data dict` attributes—which we used to invoke methods like `.keys()`—and now we don't need that functionality either.

The Python standard library provides classes similar to `Record`, where each instance has an arbitrary set of attributes built from keyword arguments given to `__init__`: `types.SimpleNamespace` (*https://fpy.li/22-5*), `argparse.Namespace` (*https://fpy.li/22-6*), and `multiprocessing.managers.Namespace` (*https://fpy.li/22-7*). I wrote the simpler `Record` class to highlight the essential idea: `__init__` updating the instance `__dict__`.

After reorganizing the schedule dataset, we can enhance the `Record` class to automatically retrieve `venue` and `speaker` records referenced in an `event` record. We'll use properties to do that in the next examples.

Step 2: Property to Retrieve a Linked Record

The goal of this next version is: given an `event` record, reading its `venue` property will return a `Record`. This is similar to what the Django ORM does when you access a `ForeignKey` field: instead of the key, you get the linked model object.

---

7 By the way, Bunch is the name of the class used by Alex Martelli to share this tip in a recipe from 2001 titled "The simple but handy 'collector of a bunch of named stuff' class" (*https://fpy.li/22-4*).

We'll start with the venue property. See the partial interaction in Example 22-10 as an example.

Example 22-10. Extract from the doctests of schedule_v2.py

```
>>> event = Record.fetch('event.33950')  ❶
>>> event  ❷
<Event 'There *Will* Be Bugs'>
>>> event.venue  ❸
<Record serial=1449>
>>> event.venue.name  ❹
'Portland 251'
>>> event.venue_serial  ❺
1449
```

❶ The Record.fetch static method gets a Record or an Event from the dataset.

❷ Note that event is an instance of the Event class.

❸ Accessing event.venue returns a Record instance.

❹ Now it's easy to find out the name of an event.venue.

❺ The Event instance also has a venue_serial attribute, from the JSON data.

Event is a subclass of Record adding a venue to retrieve linked records, and a specialized __repr__ method.

The code for this section is in the *schedule_v2.py* (*https://fpy.li/22-8*) module in the *Fluent Python* code repository (*https://fpy.li/code*). The example has nearly 60 lines, so I'll present it in parts, starting with the enhanced Record class.

Example 22-11. schedule_v2.py: Record class with a new fetch method

```
import inspect  ❶
import json

JSON_PATH = 'data/osconfeed.json'

class Record:

    __index = None  ❷

    def __init__(self, **kwargs):
        self.__dict__.update(kwargs)

    def __repr__(self):
        return f'<{self.__class__.__name__} serial={self.serial!r}>'
```

```
    @staticmethod  ❸
    def fetch(key):
        if Record.__index is None:  ❹
            Record.__index = load()
        return Record.__index[key]  ❺
```

❶ inspect will be used in load, listed in Example 22-13.

❷ The __index private class attribute will eventually hold a reference to the dict returned by load.

❸ fetch is a staticmethod to make it explicit that its effect is not influenced by the instance or class on which it is called.

❹ Populate the Record.__index, if needed.

❺ Use it to retrieve the record with the given key.

 This is one example where the use of staticmethod makes sense. The fetch method always acts on the Record.__index class attribute, even if invoked from a subclass, like Event.fetch()— which we'll soon explore. It would be misleading to code it as a class method because the cls first argument would not be used.

Now we get to the use of a property in the Event class, listed in Example 22-12.

Example 22-12. schedule_v2.py: the Event class

```
class Event(Record):  ❶

    def __repr__(self):
        try:
            return f'<{self.__class__.__name__} {self.name!r}>'  ❷
        except AttributeError:
            return super().__repr__()

    @property
    def venue(self):
        key = f'venue.{self.venue_serial}'
        return self.__class__.fetch(key)  ❸
```

❶ Event extends Record.

❷ If the instance has a name attribute, it is used to produce a custom representation. Otherwise, delegate to the __repr__ from Record.

❸ The venue property builds a key from the venue_serial attribute, and passes it to the fetch class method, inherited from Record (the reason for using self.__class__ is explained shortly).

The second line of the venue method of Example 22-12 returns self.__class__.fetch(key). Why not simply call self.fetch(key)? The simpler form works with the specific OSCON dataset because there is no event record with a 'fetch' key. But, if an event record had a key named 'fetch', then within that specific Event instance, the reference self.fetch would retrieve the value of that field, instead of the fetch class method that Event inherits from Record. This is a subtle bug, and it could easily sneak through testing because it depends on the dataset.

 When creating instance attribute names from data, there is always the risk of bugs due to shadowing of class attributes—such as methods—or data loss through accidental overwriting of existing instance attributes. These problems may explain why Python dicts are not like JavaScript objects in the first place.

If the Record class behaved more like a mapping, implementing a dynamic __getitem__ instead of a dynamic __getattr__, there would be no risk of bugs from overwriting or shadowing. A custom mapping is probably the Pythonic way to implement Record. But if I took that road, we'd not be studying the tricks and traps of dynamic attribute programming.

The final piece of this example is the revised load function in Example 22-13.

Example 22-13. schedule_v2.py: the load function

```
def load(path=JSON_PATH):
    records = {}
    with open(path) as fp:
        raw_data = json.load(fp)
    for collection, raw_records in raw_data['Schedule'].items():
        record_type = collection[:-1]          ❶
        cls_name = record_type.capitalize()     ❷
        cls = globals().get(cls_name, Record)   ❸
        if inspect.isclass(cls) and issubclass(cls, Record):   ❹
            factory = cls                       ❺
        else:
            factory = Record                    ❻
```

```
            for raw_record in raw_records:  ❼
                key = f'{record_type}.{raw_record["serial"]}'
                records[key] = factory(**raw_record)  ❽
        return records
```

❶ So far, no changes from the load in *schedule_v1.py* (Example 22-9).

❷ Capitalize the record_type to get a possible class name; e.g., 'event' becomes 'Event'.

❸ Get an object by that name from the module global scope; get the Record class if there's no such object.

❹ If the object just retrieved is a class, and is a subclass of Record…

❺ …bind the factory name to it. This means factory may be any subclass of Record, depending on the record_type.

❻ Otherwise, bind the factory name to Record.

❼ The for loop that creates the key and saves the records is the same as before, except that…

❽ …the object stored in records is constructed by factory, which may be Record or a subclass like Event, selected according to the record_type.

Note that the only record_type that has a custom class is Event, but if classes named Speaker or Venue are coded, load will automatically use those classes when building and saving records, instead of the default Record class.

We'll now apply the same idea to a new speakers property in the Events class.

Step 3: Property Overriding an Existing Attribute

The name of the venue property in Example 22-12 does not match a field name in records of the "events" collection. Its data comes from a venue_serial field name. In contrast, each record in the events collection has a speakers field with a list of serial numbers. We want to expose that information as a speakers property in Event instances, which returns a list of Record instances. This name clash requires some special attention, as Example 22-14 reveals.

Example 22-14. schedule_v3.py: the speakers property

```
@property
def speakers(self):
```

```
spkr_serials = self.__dict__['speakers']  ❶
fetch = self.__class__.fetch
return [fetch(f'speaker.{key}')
        for key in spkr_serials]  ❷
```

❶ The data we want is in a `speakers` attribute, but we must retrieve it directly from the instance `__dict__` to avoid a recursive call to the `speakers` property.

❷ Return a list of all records with keys corresponding to the numbers in `spkr_serials`.

Inside the `speakers` method, trying to read `self.speakers` will invoke the property itself, quickly raising a `RecursionError`. However, if we read the same data via `self.__dict__['speakers']`, Python's usual algorithm for retrieving attributes is bypassed, the property is not called, and the recursion is avoided. For this reason, reading or writing data directly to an object's `__dict__` is a common Python metaprogramming trick.

The interpreter evaluates `obj.my_attr` by first looking at the class of `obj`. If the class has a property with the `my_attr` name, that property shadows an instance attribute by the same name. Examples in "Properties Override Instance Attributes" on page 865 will demonstrate this, and Chapter 23 will reveal that a property is implemented as a descriptor—a more powerful and general abstraction.

As I coded the list comprehension in Example 22-14, my programmer's lizard brain thought: "This may be expensive." Not really, because events in the OSCON dataset have few speakers, so coding anything more complicated would be premature optimization. However, caching a property is a common need—and there are caveats. So let's see how to do that in the next examples.

Step 4: Bespoke Property Cache

Caching properties is a common need because there is an expectation that an expression like `event.venue` should be inexpensive.[8] Some form of caching could become necessary if the `Record.fetch` method behind the `Event` properties needed to query a database or a web API.

---

8 This is actually a downside of Meyer's Uniform Access Principle, which I mentioned in the opening of this chapter. Read the optional "Soapbox" on page 879 if you're interested in this discussion.

In the first edition *Fluent Python*, I coded the custom caching logic for the `speakers` method, as shown in Example 22-15.

Example 22-15. Custom caching logic using `hasattr` disables key-sharing optimization

```
@property
def speakers(self):
    if not hasattr(self, '__speaker_objs'):  ❶
        spkr_serials = self.__dict__['speakers']
        fetch = self.__class__.fetch
        self.__speaker_objs = [fetch(f'speaker.{key}')
                for key in spkr_serials]
    return self.__speaker_objs  ❷
```

❶ If the instance doesn't have an attribute named __speaker_objs, fetch the speaker objects and store them there.

❷ Return `self.__speaker_objs`.

The handmade caching in Example 22-15 is straightforward, but creating an attribute after the instance is initialized defeats the PEP 412—Key-Sharing Dictionary (*https://fpy.li/pep412*) optimization, as explained in "Practical Consequences of How dict Works" on page 102. Depending on the size of the dataset, the difference in memory usage may be important.

A similar hand-rolled solution that works well with the key-sharing optimization requires coding an `__init__` for the `Event` class, to create the necessary __speaker_objs initialized to `None`, and then checking for that in the `speakers` method. See Example 22-16.

Example 22-16. Storage defined in `__init__` to leverage key-sharing optimization

```
class Event(Record):

    def __init__(self, **kwargs):
        self.__speaker_objs = None
        super().__init__(**kwargs)

# 15 lines omitted...
    @property
    def speakers(self):
        if self.__speaker_objs is None:
            spkr_serials = self.__dict__['speakers']
            fetch = self.__class__.fetch
            self.__speaker_objs = [fetch(f'speaker.{key}')
                    for key in spkr_serials]
        return self.__speaker_objs
```

Examples 22-15 and 22-16 illustrate simple caching techniques that are fairly common in legacy Python codebases. However, in multithreaded programs, handmade caches like those introduce race conditions that may lead to corrupted data. If two threads are reading a property that was not previously cached, the first thread will need to compute the data for the cache attribute (`__speaker_objs` in the examples) and the second thread may read a cached value that is not yet complete.

Fortunately, Python 3.8 introduced the `@functools.cached_property` decorator, which is thread safe. Unfortunately, it comes with a couple of caveats, explained next.

Step 5: Caching Properties with functools

The `functools` module provides three decorators for caching. We saw `@cache` and `@lru_cache` in "Memoization with functools.cache" on page 322 (Chapter 9). Python 3.8 introduced `@cached_property`.

The `functools.cached_property` decorator caches the result of the method in an instance attribute with the same name. For example, in Example 22-17, the value computed by the `venue` method is stored in a `venue` attribute in `self`. After that, when client code tries to read `venue`, the newly created `venue` instance attribute is used instead of the method.

Example 22-17. Simple use of a @cached_property

```
@cached_property
def venue(self):
    key = f'venue.{self.venue_serial}'
    return self.__class__.fetch(key)
```

In "Step 3: Property Overriding an Existing Attribute" on page 856, we saw that a property shadows an instance attribute by the same name. If that is true, how can `@cached_property` work? If the property overrides the instance attribute, the `venue` attribute will be ignored and the `venue` method will always be called, computing the key and running `fetch` every time!

The answer is a bit sad: `cached_property` is a misnomer. The `@cached_property` decorator does not create a full-fledged property, it creates a *nonoverriding descriptor*. A descriptor is an object that manages the access to an attribute in another class. We will dive into descriptors in Chapter 23. The `property` decorator is a high-level API to create an *overriding descriptor*. Chapter 23 will include a through explanation about *overriding* versus *nonoverriding* descriptors.

For now, let us set aside the underlying implementation and focus on the differences between `cached_property` and `property` from a user's point of view. Raymond Hettinger explains them very well in the Python docs (*https://fpy.li/22-9*):

The mechanics of cached_property() are somewhat different from property(). A regular property blocks attribute writes unless a setter is defined. In contrast, a cached_property allows writes.

The cached_property decorator only runs on lookups and only when an attribute of the same name doesn't exist. When it does run, the cached_property writes to the attribute with the same name. Subsequent attribute reads and writes take precedence over the cached_property method and it works like a normal attribute.

The cached value can be cleared by deleting the attribute. This allows the cached_property method to run again.[9]

Back to our Event class: the specific behavior of @cached_property makes it unsuitable to decorate speakers, because that method relies on an existing attribute also named speakers, containing the serial numbers of the event speakers.

 @cached_property has some important limitations:

- It cannot be used as a drop-in replacement to @property if the decorated method already depends on an instance attribute with the same name.
- It cannot be used in a class that defines __slots__.
- It defeats the key-sharing optimization of the instance __dict__, because it creates an instance attribute after __init__.

Despite these limitations, @cached_property addresses a common need in a simple way, and it is thread safe. Its Python code (*https://fpy.li/22-13*) is an example of using a *reentrant lock* (*https://fpy.li/22-14*).

The @cached_property documentation (*https://fpy.li/22-15*) recommends an alternative solution that we can use with speakers: stacking @property and @cache decorators, as shown in Example 22-18.

Example 22-18. Stacking @property on @cache

```
@property    ❶
@cache    ❷
def speakers(self):
```

---

9 Source: @functools.cached_property (*https://fpy.li/22-9*) documentation. I know Raymond Hettinger authored this explanation because he wrote it as a response to an issue I filed: bpo42781—functools.cached_property docs should explain that it is non-overriding (*https://fpy.li/22-11*). Hettinger is a major contributor to the official Python docs and standard library. He also wrote the excellent "Descriptor HowTo Guide" (*https://fpy.li/22-12*), a key resource for Chapter 23.

```
        spkr_serials = self.__dict__['speakers']
        fetch = self.__class__.fetch
        return [fetch(f'speaker.{key}')
                for key in spkr_serials]
```

❶ The order is important: @property goes on top…

❷ …of @cache.

Recall from "Stacked Decorators" on page 324 the meaning of that syntax. The top three lines of Example 22-18 are similar to:

```
speakers = property(cache(speakers))
```

The @cache is applied to speakers, returning a new function. That function then is decorated by @property, which replaces it with a newly constructed property.

This wraps up our discussion of read-only properties and caching decorators, exploring the OSCON dataset. In the next section, we start a new series of examples creating read/write properties.

Using a Property for Attribute Validation

Besides computing attribute values, properties are also used to enforce business rules by changing a public attribute into an attribute protected by a getter and setter without affecting client code. Let's work through an extended example.

LineItem Take #1: Class for an Item in an Order

Imagine an app for a store that sells organic food in bulk, where customers can order nuts, dried fruit, or cereals by weight. In that system, each order would hold a sequence of line items, and each line item could be represented by an instance of a class, as in Example 22-19.

Example 22-19. bulkfood_v1.py: the simplest LineItem class

```
class LineItem:

    def __init__(self, description, weight, price):
        self.description = description
        self.weight = weight
        self.price = price

    def subtotal(self):
        return self.weight * self.price
```

That's nice and simple. Perhaps too simple. Example 22-20 shows a problem.

Example 22-20. A negative weight results in a negative subtotal

```
>>> raisins = LineItem('Golden raisins', 10, 6.95)
>>> raisins.subtotal()
69.5
>>> raisins.weight = -20  # garbage in...
>>> raisins.subtotal()    # garbage out...
-139.0
```

This is a toy example, but not as fanciful as you may think. Here is a story from the early days of Amazon.com:

> We found that customers could order a negative quantity of books! And we would credit their credit card with the price and, I assume, wait around for them to ship the books.
>
> — Jeff Bezos, founder and CEO of Amazon.com[10]

How do we fix this? We could change the interface of LineItem to use a getter and a setter for the weight attribute. That would be the Java way, and it's not wrong.

On the other hand, it's natural to be able to set the weight of an item by just assigning to it; and perhaps the system is in production with other parts already accessing item.weight directly. In this case, the Python way would be to replace the data attribute with a property.

LineItem Take #2: A Validating Property

Implementing a property will allow us to use a getter and a setter, but the interface of LineItem will not change (i.e., setting the weight of a LineItem will still be written as raisins.weight = 12).

Example 22-21 lists the code for a read/write weight property.

Example 22-21. bulkfood_v2.py: a LineItem with a weight property

```
class LineItem:

    def __init__(self, description, weight, price):
        self.description = description
        self.weight = weight    ❶
        self.price = price

    def subtotal(self):
        return self.weight * self.price
```

---

10 Direct quote by Jeff Bezos in the *Wall Street Journal* story, "Birth of a Salesman" (*https://fpy.li/22-16*) (October 15, 2011). Note that as of 2021, you need a subscription to read the article.

```
@property  ❷
def weight(self):  ❸
    return self.__weight  ❹

@weight.setter  ❺
def weight(self, value):
    if value > 0:
        self.__weight = value  ❻
    else:
        raise ValueError('value must be > 0')  ❼
```

❶ Here the property setter is already in use, making sure that no instances with negative weight can be created.

❷ @property decorates the getter method.

❸ All the methods that implement a property share the name of the public attribute: weight.

❹ The actual value is stored in a private attribute __weight.

❺ The decorated getter has a .setter attribute, which is also a decorator; this ties the getter and setter together.

❻ If the value is greater than zero, we set the private __weight.

❼ Otherwise, ValueError is raised.

Note how a LineItem with an invalid weight cannot be created now:

```
>>> walnuts = LineItem('walnuts', 0, 10.00)
Traceback (most recent call last):
    ...
ValueError: value must be > 0
```

Now we have protected weight from users providing negative values. Although buyers usually can't set the price of an item, a clerical error or a bug may create a LineItem with a negative price. To prevent that, we could also turn price into a property, but this would entail some repetition in our code.

Remember the Paul Graham quote from Chapter 17: "When I see patterns in my programs, I consider it a sign of trouble." The cure for repetition is abstraction. There are two ways to abstract away property definitions: using a property factory or a descriptor class. The descriptor class approach is more flexible, and we'll devote Chapter 23 to a full discussion of it. Properties are in fact implemented as descriptor classes themselves. But here we will continue our exploration of properties by implementing a property factory as a function.

But before we can implement a property factory, we need to have a deeper understanding of properties.

A Proper Look at Properties

Although often used as a decorator, the property built-in is actually a class. In Python, functions and classes are often interchangeable, because both are callable and there is no new operator for object instantiation, so invoking a constructor is no different from invoking a factory function. And both can be used as decorators, as long as they return a new callable that is a suitable replacement of the decorated callable.

This is the full signature of the property constructor:

```
property(fget=None, fset=None, fdel=None, doc=None)
```

All arguments are optional, and if a function is not provided for one of them, the corresponding operation is not allowed by the resulting property object.

The property type was added in Python 2.2, but the @ decorator syntax appeared only in Python 2.4, so for a few years, properties were defined by passing the accessor functions as the first two arguments.

The "classic" syntax for defining properties without decorators is illustrated in Example 22-22.

Example 22-22. bulkfood_v2b.py: same as Example 22-21, but without using decorators

```
class LineItem:

    def __init__(self, description, weight, price):
        self.description = description
        self.weight = weight
        self.price = price

    def subtotal(self):
        return self.weight * self.price

    def get_weight(self):        ❶
        return self.__weight

    def set_weight(self, value):        ❷
        if value > 0:
            self.__weight = value
        else:
            raise ValueError('value must be > 0')
```

```
    weight = property(get_weight, set_weight)  ❸
```

❶ A plain getter.

❷ A plain setter.

❸ Build the `property` and assign it to a public class attribute.

The classic form is better than the decorator syntax in some situations; the code of the property factory we'll discuss shortly is one example. On the other hand, in a class body with many methods, the decorators make it explicit which are the getters and setters, without depending on the convention of using `get` and `set` prefixes in their names.

The presence of a property in a class affects how attributes in instances of that class can be found in a way that may be surprising at first. The next section explains.

Properties Override Instance Attributes

Properties are always class attributes, but they actually manage attribute access in the instances of the class.

In "Overriding Class Attributes" on page 391 we saw that when an instance and its class both have a data attribute by the same name, the instance attribute overrides, or shadows, the class attribute—at least when read through that instance. Example 22-23 illustrates this point.

Example 22-23. Instance attribute shadows the class data attribute

```
>>> class Class:  ❶
...     data = 'the class data attr'
...     @property
...     def prop(self):
...         return 'the prop value'
...
>>> obj = Class()
>>> vars(obj)  ❷
{}
>>> obj.data  ❸
'the class data attr'
>>> obj.data = 'bar'  ❹
>>> vars(obj)  ❺
{'data': 'bar'}
>>> obj.data  ❻
'bar'
>>> Class.data  ❼
'the class data attr'
```

❶ Define Class with two class attributes: the data attribute and the prop property.

❷ vars returns the __dict__ of obj, showing it has no instance attributes.

❸ Reading from obj.data retrieves the value of Class.data.

❹ Writing to obj.data creates an instance attribute.

❺ Inspect the instance to see the instance attribute.

❻ Now reading from obj.data retrieves the value of the instance attribute. When read from the obj instance, the instance data shadows the class data.

❼ The Class.data attribute is intact.

Now, let's try to override the prop attribute on the obj instance. Resuming the previous console session, we have Example 22-24.

Example 22-24. Instance attribute does not shadow the class property (continued from Example 22-23)

```
>>> Class.prop  ❶
<property object at 0x1072b7408>
>>> obj.prop  ❷
'the prop value'
>>> obj.prop = 'foo'  ❸
Traceback (most recent call last):
  ...
AttributeError: can't set attribute
>>> obj.__dict__['prop'] = 'foo'  ❹
>>> vars(obj)  ❺
{'data': 'bar', 'prop': 'foo'}
>>> obj.prop  ❻
'the prop value'
>>> Class.prop = 'baz'  ❼
>>> obj.prop  ❽
'foo'
```

❶ Reading prop directly from Class retrieves the property object itself, without running its getter method.

❷ Reading obj.prop executes the property getter.

❸ Trying to set an instance prop attribute fails.

❹ Putting 'prop' directly in the obj.__dict__ works.

❺ We can see that obj now has two instance attributes: data and prop.

❻ However, reading obj.prop still runs the property getter. The property is not shadowed by an instance attribute.

❼ Overwriting Class.prop destroys the property object.

❽ Now obj.prop retrieves the instance attribute. Class.prop is not a property anymore, so it no longer overrides obj.prop.

As a final demonstration, we'll add a new property to Class, and see it overriding an instance attribute. Example 22-25 picks up where Example 22-24 left off.

Example 22-25. New class property shadows the existing instance attribute (continued from Example 22-24)

```
>>> obj.data  ❶
'bar'
>>> Class.data  ❷
'the class data attr'
>>> Class.data = property(lambda self: 'the "data" prop value')  ❸
>>> obj.data  ❹
'the "data" prop value'
>>> del Class.data  ❺
>>> obj.data  ❻
'bar'
```

❶ obj.data retrieves the instance data attribute.

❷ Class.data retrieves the class data attribute.

❸ Overwrite Class.data with a new property.

❹ obj.data is now shadowed by the Class.data property.

❺ Delete the property.

❻ obj.data now reads the instance data attribute again.

The main point of this section is that an expression like obj.data does not start the search for data in obj. The search actually starts at obj.__class__, and only if there is no property named data in the class, Python looks in the obj instance itself. This applies to *overriding descriptors* in general, of which properties are just one example. Further treatment of descriptors must wait for Chapter 23.

Now back to properties. Every Python code unit—modules, functions, classes, methods—can have a docstring. The next topic is how to attach documentation to properties.

Property Documentation

When tools such as the console help() function or IDEs need to display the documentation of a property, they extract the information from the __doc__ attribute of the property.

If used with the classic call syntax, property can get the documentation string as the doc argument:

```
weight = property(get_weight, set_weight, doc='weight in kilograms')
```

The docstring of the getter method—the one with the @property decorator itself—is used as the documentation of the property as a whole. Figure 22-1 shows the help screens generated from the code in Example 22-26.

Figure 22-1. Screenshots of the Python console when issuing the commands help(Foo.bar) and help(Foo). Source code is in Example 22-26.

Example 22-26. Documentation for a property

```
class Foo:

    @property
    def bar(self):
        """The bar attribute"""
        return self.__dict__['bar']

    @bar.setter
    def bar(self, value):
        self.__dict__['bar'] = value
```

Now that we have these property essentials covered, let's go back to the issue of protecting both the `weight` and `price` attributes of `LineItem` so they only accept values greater than zero—but without implementing two nearly identical pairs of getters/setters by hand.

Coding a Property Factory

We'll create a factory to create `quantity` properties—so named because the managed attributes represent quantities that can't be negative or zero in the application. Example 22-27 shows the clean look of the `LineItem` class using two instances of `quantity` properties: one for managing the `weight` attribute, the other for `price`.

Example 22-27. bulkfood_v2prop.py: the `quantity` property factory in use

```
class LineItem:
    weight = quantity('weight')    ❶
    price = quantity('price')      ❷

    def __init__(self, description, weight, price):
        self.description = description
        self.weight = weight       ❸
        self.price = price

    def subtotal(self):
        return self.weight * self.price    ❹
```

❶ Use the factory to define the first custom property, `weight`, as a class attribute.

❷ This second call builds another custom property, `price`.

❸ Here the property is already active, making sure a negative or 0 `weight` is rejected.

❹ The properties are also in use here, retrieving the values stored in the instance.

Recall that properties are class attributes. When building each `quantity` property, we need to pass the name of the `LineItem` attribute that will be managed by that specific property. Having to type the word `weight` twice in this line is unfortunate:

```
    weight = quantity('weight')
```

But avoiding that repetition is complicated because the property has no way of knowing which class attribute name will be bound to it. Remember: the righthand side of an assignment is evaluated first, so when `quantity()` is invoked, the `weight` class attribute doesn't even exist.

Improving the quantity property so that the user doesn't need to retype the attribute name is a nontrivial metaprogramming problem. We'll solve that problem in Chapter 23.

Example 22-28 lists the implementation of the quantity property factory.[11]

Example 22-28. bulkfood_v2prop.py: the quantity property factory

```
def quantity(storage_name):    ❶

    def qty_getter(instance):    ❷
        return instance.__dict__[storage_name]    ❸

    def qty_setter(instance, value):    ❹
        if value > 0:
            instance.__dict__[storage_name] = value    ❺
        else:
            raise ValueError('value must be > 0')

    return property(qty_getter, qty_setter)    ❻
```

❶ The storage_name argument determines where the data for each property is stored; for the weight, the storage name will be 'weight'.

❷ The first argument of the qty_getter could be named self, but that would be strange because this is not a class body; instance refers to the LineItem instance where the attribute will be stored.

❸ qty_getter references storage_name, so it will be preserved in the closure of this function; the value is retrieved directly from the instance.__dict__ to bypass the property and avoid an infinite recursion.

❹ qty_setter is defined, also taking instance as first argument.

❺ The value is stored directly in the instance.__dict__, again bypassing the property.

❻ Build a custom property object and return it.

---

11 This code is adapted from "Recipe 9.21. Avoiding Repetitive Property Methods" from *Python Cookbook*, 3rd ed., by David Beazley and Brian K. Jones (O'Reilly).

The bits of Example 22-28 that deserve careful study revolve around the stor age_name variable. When you code each property in the traditional way, the name of the attribute where you will store a value is hardcoded in the getter and setter methods. But here, the qty_getter and qty_setter functions are generic, and they depend on the storage_name variable to know where to get/set the managed attribute in the instance __dict__. Each time the quantity factory is called to build a property, the storage_name must be set to a unique value.

The functions qty_getter and qty_setter will be wrapped by the property object created in the last line of the factory function. Later, when called to perform their duties, these functions will read the storage_name from their closures to determine where to retrieve/store the managed attribute values.

In Example 22-29, I create and inspect a LineItem instance, exposing the storage attributes.

Example 22-29. bulkfood_v2prop.py: exploring properties and storage attributes

```
>>> nutmeg = LineItem('Moluccan nutmeg', 8, 13.95)
>>> nutmeg.weight, nutmeg.price  ❶
(8, 13.95)
>>> nutmeg.__dict__  ❷
{'description': 'Moluccan nutmeg', 'weight': 8, 'price': 13.95}
```

❶ Reading the weight and price through the properties shadowing the namesake instance attributes.

❷ Using vars to inspect the nutmeg instance: here we see the actual instance attributes used to store the values.

Note how the properties built by our factory leverage the behavior described in "Properties Override Instance Attributes" on page 865: the weight property overrides the weight instance attribute so that every reference to self.weight or nut meg.weight is handled by the property functions, and the only way to bypass the property logic is to access the instance __dict__ directly.

The code in Example 22-28 may be a bit tricky, but it's concise: it's identical in length to the decorated getter/setter pair defining just the weight property in Example 22-21. The LineItem definition in Example 22-27 looks much better without the noise of the getter/setters.

In a real system, that same kind of validation may appear in many fields, across several classes, and the quantity factory would be placed in a utility module to be used over and over again. Eventually that simple factory could be refactored into a more

extensible descriptor class, with specialized subclasses performing different valida-
tions. We'll do that in Chapter 23.

Now let us wrap up the discussion of properties with the issue of attribute deletion.

Handling Attribute Deletion

We can use the del statement to delete not only variables, but also attributes:

```
>>> class Demo:
...     pass
...
>>> d = Demo()
>>> d.color = 'green'
>>> d.color
'green'
>>> del d.color
>>> d.color
Traceback (most recent call last):
  File "<stdin>", line 1, in <module>
AttributeError: 'Demo' object has no attribute 'color'
```

In practice, deleting attributes is not something we do every day in Python, and the
requirement to handle it with a property is even more unusual. But it is supported,
and I can think of a silly example to demonstrate it.

In a property definition, the @my_property.deleter decorator wraps the method in
charge of deleting the attribute managed by the property. As promised, silly
Example 22-30 is inspired by the scene with the Black Knight from *Monty Python
and the Holy Grail*.[12]

Example 22-30. blackknight.py

```
class BlackKnight:

    def __init__(self):
        self.phrases = [
            ('an arm', "'Tis but a scratch."),
            ('another arm', "It's just a flesh wound."),
            ('a leg', "I'm invincible!"),
            ('another leg', "All right, we'll call it a draw.")
        ]

    @property
    def member(self):
        print('next member is:')
```

---

12 The bloody scene is available on Youtube (*https://fpy.li/22-17*) as I review this in October 2021.

```
    return self.phrases[0][0]

@member.deleter
def member(self):
    member, text = self.phrases.pop(0)
    print(f'BLACK KNIGHT (loses {member}) -- {text}')
```

The doctests in *blackknight.py* are in Example 22-31.

Example 22-31. blackknight.py: doctests for Example 22-30 (the Black Knight never concedes defeat)

```
>>> knight = BlackKnight()
>>> knight.member
next member is:
'an arm'
>>> del knight.member
BLACK KNIGHT (loses an arm) -- 'Tis but a scratch.
>>> del knight.member
BLACK KNIGHT (loses another arm) -- It's just a flesh wound.
>>> del knight.member
BLACK KNIGHT (loses a leg) -- I'm invincible!
>>> del knight.member
BLACK KNIGHT (loses another leg) -- All right, we'll call it a draw.
```

Using the classic call syntax instead of decorators, the fdel argument configures the deleter function. For example, the member property would be coded like this in the body of the BlackKnight class:

```
member = property(member_getter, fdel=member_deleter)
```

If you are not using a property, attribute deletion can also be handled by implementing the lower-level __delattr__ special method, presented in "Special Methods for Attribute Handling" on page 875. Coding a silly class with __delattr__ is left as an exercise to the procrastinating reader.

Properties are a powerful feature, but sometimes simpler or lower-level alternatives are preferable. In the final section of this chapter, we'll review some of the core APIs that Python offers for dynamic attribute programming.

Essential Attributes and Functions for Attribute Handling

Throughout this chapter, and even before in the book, we've used some of the built-in functions and special methods Python provides for dealing with dynamic attributes. This section gives an overview of them in one place, because their documentation is scattered in the official docs.

Special Attributes that Affect Attribute Handling

The behavior of many of the functions and special methods listed in the following sections depend on three special attributes:

__class__
> A reference to the object's class (i.e., obj.__class__ is the same as type(obj)). Python looks for special methods such as __getattr__ only in an object's class, and not in the instances themselves.

__dict__
> A mapping that stores the writable attributes of an object or class. An object that has a __dict__ can have arbitrary new attributes set at any time. If a class has a __slots__ attribute, then its instances may not have a __dict__. See __slots__ (next).

__slots__
> An attribute that may be defined in a class to save memory. __slots__ is a tuple of strings naming the allowed attributes.[13] If the '__dict__' name is not in __slots__, then the instances of that class will not have a __dict__ of their own, and only the attributes listed in __slots__ will be allowed in those instances. Recall "Saving Memory with __slots__" on page 386 for more.

Built-In Functions for Attribute Handling

These five built-in functions perform object attribute reading, writing, and introspection:

dir([object])
> Lists most attributes of the object. The official docs (*https://fpy.li/22-18*) say dir is intended for interactive use so it does not provide a comprehensive list of attributes, but an "interesting" set of names. dir can inspect objects implemented with or without a __dict__. The __dict__ attribute itself is not listed by dir, but the __dict__ keys are listed. Several special attributes of classes, such as __mro__, __bases__, and __name__, are not listed by dir either. You can customize the output of dir by implementing the __dir__ special method, as we saw in Example 22-4. If the optional object argument is not given, dir lists the names in the current scope.

---

13 Alex Martelli points out that, although __slots__ can be coded as a list, it's better to be explicit and always use a tuple, because changing the list in the __slots__ after the class body is processed has no effect, so it would be misleading to use a mutable sequence there.

```
getattr(object, name[, default])
```
Gets the attribute identified by the name string from the object. The main use case is to retrieve attributes (or methods) whose names we don't know beforehand. This may fetch an attribute from the object's class or from a superclass. If no such attribute exists, getattr raises AttributeError or returns the default value, if given. One great example of using gettatr is in the Cmd.onecmd method (*https://fpy.li/22-19*) in the cmd package of the standard library, where it is used to get and execute a user-defined command.

```
hasattr(object, name)
```
Returns True if the named attribute exists in the object, or can be somehow fetched through it (by inheritance, for example). The documentation (*https://fpy.li/22-20*) explains: "This is implemented by calling getattr(object, name) and seeing whether it raises an AttributeError or not."

```
setattr(object, name, value)
```
Assigns the value to the named attribute of object, if the object allows it. This may create a new attribute or overwrite an existing one.

```
vars([object])
```
Returns the __dict__ of object; vars can't deal with instances of classes that define __slots__ and don't have a __dict__ (contrast with dir, which handles such instances). Without an argument, vars() does the same as locals(): returns a dict representing the local scope.

Special Methods for Attribute Handling

When implemented in a user-defined class, the special methods listed here handle attribute retrieval, setting, deletion, and listing.

Attribute access using either dot notation or the built-in functions getattr, hasattr, and setattr triggers the appropriate special methods listed here. Reading and writing attributes directly in the instance __dict__ does not trigger these special methods —and that's the usual way to bypass them if needed.

Section "3.3.11. Special method lookup" (*https://fpy.li/22-21*) of the "Data model" chapter warns:

> For custom classes, implicit invocations of special methods are only guaranteed to work correctly if defined on an object's type, not in the object's instance dictionary.

In other words, assume that the special methods will be retrieved on the class itself, even when the target of the action is an instance. For this reason, special methods are not shadowed by instance attributes with the same name.

In the following examples, assume there is a class named Class, obj is an instance of Class, and attr is an attribute of obj.

For every one of these special methods, it doesn't matter if the attribute access is done using dot notation or one of the built-in functions listed in "Built-In Functions for Attribute Handling" on page 874. For example, both obj.attr and getattr(obj, 'attr', 42) trigger Class.__getattribute__(obj, 'attr').

__delattr__(self, name)
> Always called when there is an attempt to delete an attribute using the del statement; e.g., del obj.attr triggers Class.__delattr__(obj, 'attr'). If attr is a property, its deleter method is never called if the class implements __delattr__.

__dir__(self)
> Called when dir is invoked on the object, to provide a listing of attributes; e.g., dir(obj) triggers Class.__dir__(obj). Also used by tab-completion in all modern Python consoles.

__getattr__(self, name)
> Called only when an attempt to retrieve the named attribute fails, after the obj, Class, and its superclasses are searched. The expressions obj.no_such_attr, getattr(obj, 'no_such_attr'), and hasattr(obj, 'no_such_attr') may trigger Class.__getattr__(obj, 'no_such_attr'), but only if an attribute by that name cannot be found in obj or in Class and its superclasses.

__getattribute__(self, name)
> Always called when there is an attempt to retrieve the named attribute directly from Python code (the interpreter may bypass this in some cases, for example, to get the __repr__ method). Dot notation and the getattr and hasattr built-ins trigger this method. __getattr__ is only invoked after __getattribute__, and only when __getattribute__ raises AttributeError. To retrieve attributes of the instance obj without triggering an infinite recursion, implementations of __getattribute__ should use super().__getattribute__(obj, name).

__setattr__(self, name, value)
> Always called when there is an attempt to set the named attribute. Dot notation and the setattr built-in trigger this method; e.g., both obj.attr = 42 and setattr(obj, 'attr', 42) trigger Class.__setattr__(obj, 'attr', 42).

In practice, because they are unconditionally called and affect practically every attribute access, the `__getattribute__` and `__setattr__` special methods are harder to use correctly than `__getattr__`, which only handles nonexisting attribute names. Using properties or descriptors is less error prone than defining these special methods.

This concludes our dive into properties, special methods, and other techniques for coding dynamic attributes.

Chapter Summary

We started our coverage of dynamic attributes by showing practical examples of simple classes to make it easier to deal with a JSON dataset. The first example was the FrozenJSON class that converted nested dicts and lists into nested FrozenJSON instances and lists of them. The FrozenJSON code demonstrated the use of the `__getattr__` special method to convert data structures on the fly, whenever their attributes were read. The last version of FrozenJSON showcased the use of the `__new__` constructor method to transform a class into a flexible factory of objects, not limited to instances of itself.

We then converted the JSON dataset to a dict storing instances of a Record class. The first rendition of Record was a few lines long and introduced the "bunch" idiom: using `self.__dict__.update(**kwargs)` to build arbitrary attributes from keyword arguments passed to `__init__`. The second iteration added the Event class, implementing automatic retrieval of linked records through properties. Computed property values sometimes require caching, and we covered a few ways of doing that.

After realizing that `@functools.cached_property` is not always applicable, we learned about an alternative: combining `@property` on top of `@functools.cache`, in that order.

Coverage of properties continued with the LineItem class, where a property was deployed to protect a weight attribute from negative or zero values that make no business sense. After a deeper look at property syntax and semantics, we created a property factory to enforce the same validation on weight and price, without coding multiple getters and setters. The property factory leveraged subtle concepts—such as closures, and instance attribute overriding by properties—to provide an elegant generic solution using the same number of lines as a single hand-coded property definition.

Finally, we had a brief look at handling attribute deletion with properties, followed by an overview of the key special attributes, built-in functions, and special methods that support attribute metaprogramming in the core Python language.

Further Reading

The official documentation for the attribute handling and introspection built-in functions is Chapter 2, "Built-in Functions" (*https://fpy.li/22-22*) of *The Python Standard Library*. The related special methods and the __slots__ special attribute are documented in *The Python Language Reference* in "3.3.2. Customizing attribute access" (*https://fpy.li/22-23*). The semantics of how special methods are invoked bypassing instances is explained in "3.3.9. Special method lookup" (*https://fpy.li/22-24*). In Chapter 4, "Built-in Types," of *The Python Standard Library*, "4.13. Special Attributes" (*https://fpy.li/22-25*) covers __class__ and __dict__ attributes.

Python Cookbook, 3rd ed., by David Beazley and Brian K. Jones (O'Reilly) has several recipes covering the topics of this chapter, but I will highlight three that are outstanding: "Recipe 8.8. Extending a Property in a Subclass" addresses the thorny issue of overriding the methods inside a property inherited from a superclass; "Recipe 8.15. Delegating Attribute Access" implements a proxy class showcasing most special methods from "Special Methods for Attribute Handling" on page 875 in this book; and the awesome "Recipe 9.21. Avoiding Repetitive Property Methods," which was the basis for the property factory function presented in Example 22-28.

Python in a Nutshell, 3rd ed., by Alex Martelli, Anna Ravenscroft, and Steve Holden (O'Reilly) is rigorous and objective. They devote only three pages to properties, but that's because the book follows an axiomatic presentation style: the preceding 15 pages or so provide a thorough description of the semantics of Python classes from the ground up, including descriptors, which are how properties are actually implemented under the hood. So by the time Martelli et al., get to properties, they pack a lot of insights in those three pages—including what I selected to open this chapter.

Bertrand Meyer—quoted in the Uniform Access Principle definition in this chapter opening—pioneered the Design by Contract methodology, designed the Eiffel language, and wrote the excellent *Object-Oriented Software Construction*, 2nd ed. (Pearson). The first six chapters provide one of the best conceptual introductions to OO analysis and design I've seen. Chapter 11 presents Design by Contract, and Chapter 35 offers Meyer's assessments of some influential object-oriented languages: Simula, Smalltalk, CLOS (the Common Lisp Object System), Objective-C, C++, and Java, with brief comments on some others. Only in the last page of the book does he reveal that the highly readable "notation" he uses as pseudocode is Eiffel.

Soapbox

Meyer's Uniform Access Principle is aesthetically appealing. As a programmer using an API, I shouldn't have to care whether `product.price` simply fetches a data attribute or performs a computation. As a consumer and a citizen, I do care: in e-commerce today the value of `product.price` often depends on who is asking, so it's certainly not a mere data attribute. In fact, it's common practice that the price is lower if the query comes from outside the store—say, from a price-comparison engine. This effectively punishes loyal customers who like to browse within a particular store. But I digress.

The previous digression does raise a relevant point for programming: although the Uniform Access Principle makes perfect sense in an ideal world, in reality, users of an API may need to know whether reading `product.price` is potentially too expensive or time-consuming. That's a problem with programming abstractions in general: they make it hard to reason about the runtime cost of evaluating an expression. On the other hand, abstractions let users accomplish more with less code. It's a trade-off. As usual in matters of software engineering, Ward Cunningham's original wiki (*https://fpy.li/22-26*) hosts insightful arguments about the merits of the Uniform Access Principle (*https://fpy.li/22-27*).

In object-oriented programming languages, application or violations of the Uniform Access Principle often revolve around the syntax of reading public data attributes versus invoking getter/setter methods.

Smalltalk and Ruby address this issue in a simple and elegant way: they don't support public data attributes at all. Every instance attribute in these languages is private, so every access to them must be through methods. But their syntax makes this painless: in Ruby, `product.price` invokes the `price` getter; in Smalltalk, it's simply `product price`.

At the other end of the spectrum, the Java language allows the programmer to choose among four access-level modifiers—including the no-name default that the Java Tutorial (*https://fpy.li/22-28*) calls "package-private."

The general practice does not agree with the syntax established by the Java designers, though. Everybody in Java-land agrees that attributes should be `private`, and you must spell it out every time, because it's not the default. When all attributes are private, all access to them from outside the class must go through accessors. Java IDEs include shortcuts for generating accessor methods automatically. Unfortunately, the IDE is not so helpful when you must read the code six months later. It's up to you to wade through a sea of do-nothing accessors to find those that add value by implementing some business logic.

Alex Martelli speaks for the majority of the Python community when he calls accessors "goofy idioms" and then provides these examples that look very different but do the same thing:[14]

```
someInstance.widgetCounter += 1
# rather than...
someInstance.setWidgetCounter(someInstance.getWidgetCounter() + 1)
```

Sometimes when designing APIs, I've wondered whether every method that does not take an argument (besides `self`), returns a value (other than `None`), and is a pure function (i.e., has no side effects) should be replaced by a read-only property. In this chapter, the `LineItem.subtotal` method (as in Example 22-27) would be a good candidate to become a read-only property. Of course, this excludes methods that are designed to change the object, such as `my_list.clear()`. It would be a terrible idea to turn that into a property, so that merely accessing `my_list.clear` would delete the contents of the list!

In the *Pingo* (*https://fpy.li/22-29*) GPIO library, which I coauthored (mentioned in "The __missing__ Method" on page 91), much of the user-level API is based on properties. For example, to read the current value of an analog pin, the user writes `pin.value`, and setting a digital pin mode is written as `pin.mode = OUT`. Behind the scenes, reading an analog pin value or setting a digital pin mode may involve a lot of code, depending on the specific board driver. We decided to use properties in Pingo because we want the API to be comfortable to use even in interactive environments like a Jupyter Notebook, and we feel `pin.mode = OUT` is easier on the eyes and on the fingers than `pin.set_mode(OUT)`.

Although I find the Smalltalk and Ruby solution cleaner, I think the Python approach makes more sense than the Java one. We are allowed to start simple, coding data members as public attributes, because we know they can always be wrapped by properties (or descriptors, which we'll talk about in the next chapter).

__new__ Is Better than new

Another example of the Uniform Access Principle (or a variation of it) is the fact that function calls and object instantiation use the same syntax in Python: `my_obj = foo()`, where `foo` may be a class or any other callable.

Other languages influenced by C++ syntax have a `new` operator that makes instantiation look different than a call. Most of the time, the user of an API doesn't care whether `foo` is a function or a class. For years I was under the impression that `property` was a function. In normal usage, it makes no difference.

---

14 Alex Martelli, *Python in a Nutshell*, 2nd ed. (O'Reilly), p. 101.

There are many good reasons for replacing constructors with factories.[15] A popular motive is limiting the number of instances by returning previously built ones (as in the Singleton pattern). A related use is caching expensive object construction. Also, sometimes it's convenient to return objects of different types, depending on the arguments given.

Coding a constructor is simpler; providing a factory adds flexibility at the expense of more code. In languages that have a new operator, the designer of an API must decide in advance whether to stick with a simple constructor or invest in a factory. If the initial choice is wrong, the correction may be costly—all because new is an operator.

Sometimes it may also be convenient to go the other way, and replace a simple function with a class.

In Python, classes and functions are interchangeable in many situations. Not only because there's no new operator, but also because there is the __new__ special method, which can turn a class into a factory producing objects of different kinds (as we saw in "Flexible Object Creation with __new__" on page 847) or returning prebuilt instances instead of creating a new one every time.

This function-class duality would be easier to leverage if PEP 8 — Style Guide for Python Code (*https://fpy.li/22-31*) did not recommend CamelCase for class names. On the other hand, dozens of classes in the standard library have lowercase names (e.g., property, str, defaultdict, etc.). So maybe the use of lowercase class names is a feature, and not a bug. But however we look at it, the inconsistent capitalization of classes in the Python standard library poses a usability problem.

Although calling a function is not different from calling a class, it's good to know which is which because of another thing we can do with a class: subclassing. So I personally use CamelCase in every class that I code, and I wish all classes in the Python standard library used the same convention. I am looking at you, collections.OrderedDict and collections.defaultdict.

---

15 The reasons I am about to mention are given in the Dr. Dobbs Journal article titled "Java's new Considered Harmful" (*https://fpy.li/22-30*), by Jonathan Amsterdam and in "Consider static factory methods instead of constructors," which is Item 1 of the award-winning book *Effective Java*, 3rd ed., by Joshua Bloch (Addison-Wesley).

CHAPTER 23

Attribute Descriptors

> Learning about descriptors not only provides access to a larger toolset, it creates a deeper understanding of how Python works and an appreciation for the elegance of its design.
>
> — Raymond Hettinger, Python core developer and guru[1]

Descriptors are a way of reusing the same access logic in multiple attributes. For example, field types in ORMs, such as the Django ORM and SQLAlchemy, are descriptors, managing the flow of data from the fields in a database record to Python object attributes and vice versa.

A descriptor is a class that implements a dynamic protocol consisting of the __get__, __set__, and __delete__ methods. The property class implements the full descriptor protocol. As usual with dynamic protocols, partial implementations are OK. In fact, most descriptors we see in real code implement only __get__ and __set__, and many implement only one of these methods.

Descriptors are a distinguishing feature of Python, deployed not only at the application level but also in the language infrastructure. User-defined functions are descriptors. We'll see how the descriptor protocol allows methods to operate as bound or unbound methods, depending on how they are called.

Understanding descriptors is key to Python mastery. This is what this chapter is about.

In this chapter we'll refactor the bulk food example we first saw in "Using a Property for Attribute Validation" on page 861, replacing properties with descriptors. This will make it easier to reuse the attribute validation logic across different classes. We'll

---

1 Raymond Hettinger, *Descriptor HowTo Guide* (*https://fpy.li/descrhow*).

tackle the concepts of overriding and nonoverriding descriptors, and realize that Python functions are descriptors. Finally we'll see some tips about implementing descriptors.

What's New in This Chapter

The Quantity descriptor example in "LineItem Take #4: Automatic Naming of Storage Attributes" on page 891 was dramatically simplified thanks to the __set_name__ special method added to the descriptor protocol in Python 3.6.

I removed the property factory example formerly in "LineItem Take #4: Automatic Naming of Storage Attributes" on page 891 because it became irrelevant: the point was to show an alternative way of solving the Quantity problem, but with the addition of __set_name__, the descriptor solution becomes much simpler.

The AutoStorage class that used to appear in "LineItem Take #5: A New Descriptor Type" on page 893 is also gone because __set_name__ made it obsolete.

Descriptor Example: Attribute Validation

As we saw in "Coding a Property Factory" on page 869, a property factory is a way to avoid repetitive coding of getters and setters by applying functional programming patterns. A property factory is a higher-order function that creates a parameterized set of accessor functions and builds a custom property instance from them, with closures to hold settings like the storage_name. The object-oriented way of solving the same problem is a descriptor class.

We'll continue the series of LineItem examples where we left off, in "Coding a Property Factory" on page 869, by refactoring the quantity property factory into a Quantity descriptor class. This will make it easier to use.

LineItem Take #3: A Simple Descriptor

As we said in the introduction, a class implementing a __get__, a __set__, or a __delete__ method is a descriptor. You use a descriptor by declaring instances of it as class attributes of another class.

We'll create a Quantity descriptor, and the LineItem class will use two instances of Quantity: one for managing the weight attribute, the other for price. A diagram helps, so take a look at Figure 23-1.

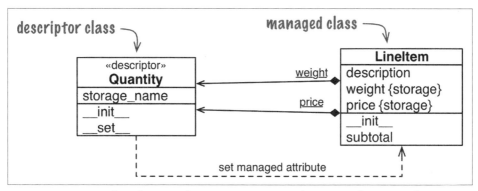

Figure 23-1. UML class diagram for `LineItem` *using a descriptor class named* `Quantity`. *Underlined attributes in UML are class attributes. Note that weight and price are instances of* `Quantity` *attached to the* `LineItem` *class, but* `LineItem` *instances also have their own weight and price attributes where those values are stored.*

Note that the word `weight` appears twice in Figure 23-1, because there are really two distinct attributes named `weight`: one is a class attribute of `LineItem`, the other is an instance attribute that will exist in each `LineItem` object. This also applies to `price`.

Terms to understand descriptors

Implementing and using descriptors involves several components, and it is useful to be precise when naming those components. I will use the following terms and definitions as I describe the examples in this chapter. They will be easier to understand once you see the code, but I wanted to put the definitions up front so you can refer back to them when needed.

Descriptor class
  A class implementing the descriptor protocol. That's `Quantity` in Figure 23-1.

Managed class
  The class where the descriptor instances are declared as class attributes. In Figure 23-1, `LineItem` is the managed class.

Descriptor instance
  Each instance of a descriptor class, declared as a class attribute of the managed class. In Figure 23-1, each descriptor instance is represented by a composition arrow with an underlined name (the underline means class attribute in UML). The black diamonds touch the `LineItem` class, which contains the descriptor instances.

Managed instance

One instance of the managed class. In this example, LineItem instances are the managed instances (they are not shown in the class diagram).

Storage attribute

An attribute of the managed instance that holds the value of a managed attribute for that particular instance. In Figure 23-1, the LineItem instance attributes weight and price are the storage attributes. They are distinct from the descriptor instances, which are always class attributes.

Managed attribute

A public attribute in the managed class that is handled by a descriptor instance, with values stored in storage attributes. In other words, a descriptor instance and a storage attribute provide the infrastructure for a managed attribute.

It's important to realize that Quantity instances are class attributes of LineItem. This crucial point is highlighted by the mills and gizmos in Figure 23-2.

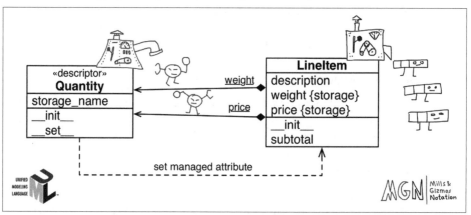

Figure 23-2. UML class diagram annotated with MGN (Mills & Gizmos Notation): classes are mills that produce gizmos—the instances. The Quantity mill produces two gizmos with round heads, which are attached to the LineItem mill: weight and price. The LineItem mill produces rectangular gizmos that have their own weight and price attributes where those values are stored.

Introducing Mills & Gizmos Notation

After explaining descriptors many times, I realized UML is not very good at showing relationships involving classes and instances, like the relationship between a managed class and the descriptor instances.[2] So I invented my own "language," the Mills & Gizmos Notation (MGN), which I use to annotate UML diagrams.

MGN is designed to make very clear the distinction between classes and instances. See Figure 23-3. In MGN, a class is drawn as a "mill," a complicated machine that produces gizmos. Classes/mills are always machines with levers and dials. The gizmos are the instances, and they look much simpler. When this book is rendered in color, gizmos have the same color as the mill that made it.

Figure 23-3. MGN sketch showing the LineItem class making three instances, and Quantity making two. One instance of Quantity is retrieving a value stored in a LineItem instance.

For this example, I drew LineItem instances as rows in a tabular invoice, with three cells representing the three attributes (description, weight, and price). Because Quantity instances are descriptors, they have a magnifying glass to __get__ values, and a claw to __set__ values. When we get to metaclasses, you'll thank me for these doodles.

Enough doodling for now. Here is the code: Example 23-1 shows the Quantity descriptor class, and Example 23-2 lists a new LineItem class using two instances of Quantity.

---

2 Classes and instances are drawn as rectangles in UML class diagrams. There are visual differences, but instances are rarely shown in class diagrams, so developers may not recognize them as such.

Example 23-1. bulkfood_v3.py: Quantity descriptor does not accept negative values

```
class Quantity:    ❶

    def __init__(self, storage_name):
        self.storage_name = storage_name    ❷

    def __set__(self, instance, value):    ❸
        if value > 0:
            instance.__dict__[self.storage_name] = value    ❹
        else:
            msg = f'{self.storage_name} must be > 0'
            raise ValueError(msg)

    def __get__(self, instance, owner):    ❺
        return instance.__dict__[self.storage_name]
```

❶ Descriptor is a protocol-based feature; no subclassing is needed to implement one.

❷ Each Quantity instance will have a storage_name attribute: that's the name of the storage attribute to hold the value in the managed instances.

❸ __set__ is called when there is an attempt to assign to the managed attribute. Here, self is the descriptor instance (i.e., LineItem.weight or LineItem.price), instance is the managed instance (a LineItem instance), and value is the value being assigned.

❹ We must store the attribute value directly into __dict__; calling set attr (instance, self.storage_name) would trigger the __set__ method again, leading to infinite recursion.

❺ We need to implement __get__ because the name of the managed attribute may not be the same as the storage_name. The owner argument will be explained shortly.

Implementing __get__ is necessary because a user could write something like this:

```
class House:
    rooms = Quantity('number_of_rooms')
```

In the House class, the managed attribute is rooms, but the storage attribute is number_of_rooms. Given a House instance named chaos_manor, reading and writing chaos_manor.rooms goes through the Quantity descriptor instance attached to rooms, but reading and writing chaos_manor.number_of_rooms bypasses the descriptor.

Note that __get__ receives three arguments: self, instance, and owner. The owner argument is a reference to the managed class (e.g., LineItem), and it's useful if you want the descriptor to support retrieving a class attribute—perhaps to emulate Python's default behavior of retrieving a class attribute when the name is not found in the instance.

If a managed attribute, such as weight, is retrieved via the class like Line Item.weight, the descriptor __get__ method receives None as the value for the instance argument.

To support introspection and other metaprogramming tricks by the user, it's a good practice to make __get__ return the descriptor instance when the managed attribute is accessed through the class. To do that, we'd code __get__ like this:

```python
def __get__(self, instance, owner):
    if instance is None:
        return self
    else:
        return instance.__dict__[self.storage_name]
```

Example 23-2 demonstrates the use of Quantity in LineItem.

Example 23-2. bulkfood_v3.py: Quantity descriptors manage attributes in LineItem

```python
class LineItem:
    weight = Quantity('weight')  ❶
    price = Quantity('price')  ❷

    def __init__(self, description, weight, price):  ❸
        self.description = description
        self.weight = weight
        self.price = price

    def subtotal(self):
        return self.weight * self.price
```

❶ The first descriptor instance will manage the weight attribute.

❷ The second descriptor instance will manage the price attribute.

❸ The rest of the class body is as simple and clean as the original code in *bulk-food_v1.py* (Example 22-19).

The code in Example 23-2 works as intended, preventing the sale of truffles for $0:[3]

```
>>> truffle = LineItem('White truffle', 100, 0)
Traceback (most recent call last):
  ...
ValueError: value must be > 0
```

 When coding descriptor __get__ and __set__ methods, keep in mind what the self and instance arguments mean: self is the descriptor instance, and instance is the managed instance. Descriptors managing instance attributes should store values in the managed instances. That's why Python provides the instance argument to the descriptor methods.

It may be tempting, but wrong, to store the value of each managed attribute in the descriptor instance itself. In other words, in the __set__ method, instead of coding:

```
instance.__dict__[self.storage_name] = value
```

the tempting, but bad, alternative would be:

```
self.__dict__[self.storage_name] = value
```

To understand why this would be wrong, think about the meaning of the first two arguments to __set__: self and instance. Here, self is the descriptor instance, which is actually a class attribute of the managed class. You may have thousands of LineItem instances in memory at one time, but you'll only have two instances of the descriptors: the class attributes LineItem.weight and LineItem.price. So anything you store in the descriptor instances themselves is actually part of a LineItem class attribute, and therefore is shared among all LineItem instances.

A drawback of Example 23-2 is the need to repeat the names of the attributes when the descriptors are instantiated in the managed class body. It would be nice if the LineItem class could be declared like this:

```
class LineItem:
    weight = Quantity()
    price = Quantity()

    # remaining methods as before
```

As it stands, Example 23-2 requires naming each Quantity explicitly, which is not only inconvenient but dangerous. If a programmer copying and pasting code forgets to edit both names and writes something like price = Quantity('weight'), the

---

3 White truffles cost thousands of dollars per pound. Disallowing the sale of truffles for $0.01 is left as an exercise for the enterprising reader. I know a person who actually bought an $1,800 encyclopedia of statistics for $18 because of an error in an online store (not *Amazon.com* in this case).

program will misbehave badly, clobbering the value of `weight` whenever the `price` is set.

The problem is that—as we saw in Chapter 6—the righthand side of an assignment is executed before the variable exists. The expression `Quantity()` is evaluated to create a descriptor instance, and there is no way the code in the `Quantity` class can guess the name of the variable to which the descriptor will be bound (e.g., `weight` or `price`).

Thankfully, the descriptor protocol now supports the aptly named __set_name__ special method. We'll see how to use it next.

 Automatic naming of a descriptor storage attribute used to be a thorny issue. In the first edition of *Fluent Python*, I devoted several pages and lines of code in this chapter and the next to presenting different solutions, including the use of a class decorator, and then metaclasses in Chapter 24. This was greatly simplified in Python 3.6.

LineItem Take #4: Automatic Naming of Storage Attributes

To avoid retyping the attribute name in the descriptor instances, we'll implement __set_name__ to set the `storage_name` of each `Quantity` instance. The __set_name__ special method was added to the descriptor protocol in Python 3.6. The interpreter calls __set_name__ on each descriptor it finds in a `class` body—if the descriptor implements it.[4]

In Example 23-3, the `LineItem` descriptor class doesn't need an __init__. Instead, __set_item__ saves the name of the storage attribute.

Example 23-3. bulkfood_v4.py: __set_name__ sets the name for each Quantity descriptor instance

```
class Quantity:

    def __set_name__(self, owner, name):    ❶
        self.storage_name = name            ❷

    def __set__(self, instance, value):     ❸
        if value > 0:
            instance.__dict__[self.storage_name] = value
```

---

4 More precisely, __set_name__ is called by `type.__new__`—the constructor of objects representing classes. The type built-in is actually a metaclass, the default class of user-defined classes. This is hard to grasp at first, but rest assured: Chapter 24 is devoted to the dynamic configuration of classes, including the concept of metaclasses.

```
        else:
            msg = f'{self.storage_name} must be > 0'
            raise ValueError(msg)

    # no __get__ needed  ❹

class LineItem:
    weight = Quantity()  ❺
    price = Quantity()

    def __init__(self, description, weight, price):
        self.description = description
        self.weight = weight
        self.price = price

    def subtotal(self):
        return self.weight * self.price
```

❶ self is the descriptor instance (not the managed instance), owner is the managed class, and name is the name of the attribute of owner to which this descriptor instance was assigned in the class body of owner.

❷ This is what the __init__ did in Example 23-1.

❸ The __set__ method here is exactly the same as in Example 23-1.

❹ Implementing __get__ is not necessary because the name of the storage attribute matches the name of the managed attribute. The expression product.price gets the price attribute directly from the LineItem instance.

❺ Now we don't need to pass the managed attribute name to the Quantity constructor. That was the goal for this version.

Looking at Example 23-3, you may think that's a lot of code just for managing a couple of attributes, but it's important to realize that the descriptor logic is now abstracted into a separate code unit: the Quantity class. Usually we do not define a descriptor in the same module where it's used, but in a separate utility module designed to be used across the application—even in many applications, if you are developing a library or framework.

With this in mind, Example 23-4 better represents the typical usage of a descriptor.

Example 23-4. bulkfood_v4c.py: LineItem definition uncluttered; the Quantity descriptor class now resides in the imported model_v4c module

```
import model_v4c as model  ❶
```

```
class LineItem:
    weight = model.Quantity()  ❷
    price = model.Quantity()

    def __init__(self, description, weight, price):
        self.description = description
        self.weight = weight
        self.price = price

    def subtotal(self):
        return self.weight * self.price
```

❶ Import the model_v4c module where Quantity is implemented.

❷ Put model.Quantity to use.

Django users will notice that Example 23-4 looks a lot like a model definition. It's no coincidence: Django model fields are descriptors.

Because descriptors are implemented as classes, we can leverage inheritance to reuse some of the code we have for new descriptors. That's what we'll do in the following section.

LineItem Take #5: A New Descriptor Type

The imaginary organic food store hits a snag: somehow a line item instance was created with a blank description, and the order could not be fulfilled. To prevent that, we'll create a new descriptor, NonBlank. As we design NonBlank, we realize it will be very much like the Quantity descriptor, except for the validation logic.

This prompts a refactoring, producing Validated, an abstract class that overrides the __set__ method, calling a validate method that must be implemented by subclasses.

We'll then rewrite Quantity, and implement NonBlank by inheriting from Validated and just coding the validate methods.

The relationship among Validated, Quantity, and NonBlank is an application of the *template method* as described in the *Design Patterns* classic:

> A template method defines an algorithm in terms of abstract operations that subclasses override to provide concrete behavior.[5]

---

5 Gamma et al., *Design Patterns: Elements of Reusable Object-Oriented Software*, p. 326.

In Example 23-5, Validated.__set__ is the template method and self.validate is the abstract operation.

Example 23-5. model_v5.py: the Validated ABC

```python
import abc

class Validated(abc.ABC):

    def __set_name__(self, owner, name):
        self.storage_name = name

    def __set__(self, instance, value):
        value = self.validate(self.storage_name, value)  ❶
        instance.__dict__[self.storage_name] = value  ❷

    @abc.abstractmethod
    def validate(self, name, value):  ❸
        """return validated value or raise ValueError"""
```

❶ __set__ delegates validation to the validate method...

❷ ...then uses the returned value to update the stored value.

❸ validate is an abstract method; this is the template method.

Alex Martelli prefers to call this design pattern *Self-Delegation*, and I agree it's a more descriptive name: the first line of __set__ self-delegates to validate.[6]

The concrete Validated subclasses in this example are Quantity and NonBlank, shown in Example 23-6.

Example 23-6. model_v5.py: Quantity and NonBlank, concrete Validated subclasses

```python
class Quantity(Validated):
    """a number greater than zero"""

    def validate(self, name, value):  ❶
        if value <= 0:
            raise ValueError(f'{name} must be > 0')
        return value


class NonBlank(Validated):
    """a string with at least one non-space character"""
```

---

6 Slide #50 of Alex Martelli's "Python Design Patterns" talk (*https://fpy.li/23-1*). Highly recommended.

```
def validate(self, name, value):
    value = value.strip()
    if not value:   ❷
        raise ValueError(f'{name} cannot be blank')
    return value   ❸
```

❶ Implementation of the template method required by the Validated.validate abstract method.

❷ If nothing is left after leading and trailing blanks are stripped, reject the value.

❸ Requiring the concrete validate methods to return the validated value gives them an opportunity to clean up, convert, or normalize the data received. In this case, value is returned without leading or trailing blanks.

Users of *model_v5.py* don't need to know all these details. What matters is that they get to use Quantity and NonBlank to automate the validation of instance attributes. See the latest LineItem class in Example 23-7.

Example 23-7. bulkfood_v5.py: LineItem using Quantity and NonBlank descriptors

```
import model_v5 as model   ❶

class LineItem:
    description = model.NonBlank()   ❷
    weight = model.Quantity()
    price = model.Quantity()

    def __init__(self, description, weight, price):
        self.description = description
        self.weight = weight
        self.price = price

    def subtotal(self):
        return self.weight * self.price
```

❶ Import the model_v5 module, giving it a friendlier name.

❷ Put model.NonBlank to use. The rest of the code is unchanged.

The LineItem examples we've seen in this chapter demonstrate a typical use of descriptors to manage data attributes. Descriptors like Quantity are called overriding descriptors because its __set__ method overrides (i.e., intercepts and overrules) the setting of an instance attribute by the same name in the managed instance. However, there are also nonoverriding descriptors. We'll explore this distinction in detail in the next section.

Overriding Versus Nonoverriding Descriptors

Recall that there is an important asymmetry in the way Python handles attributes. Reading an attribute through an instance normally returns the attribute defined in the instance, but if there is no such attribute in the instance, a class attribute will be retrieved. On the other hand, assigning to an attribute in an instance normally creates the attribute in the instance, without affecting the class at all.

This asymmetry also affects descriptors, in effect creating two broad categories of descriptors, depending on whether the __set__ method is implemented. If __set__ is present, the class is an overriding descriptor; otherwise, it is a nonoverriding descriptor. These terms will make sense as we study descriptor behaviors in the next examples.

Observing the different descriptor categories requires a few classes, so we'll use the code in Example 23-8 as our test bed for the following sections.

 Every __get__ and __set__ method in Example 23-8 calls print_args so their invocations are displayed in a readable way. Understanding print_args and the auxiliary functions cls_name and display is not important, so don't get distracted by them.

Example 23-8. descriptorkinds.py: simple classes for studying descriptor overriding behaviors

```
### auxiliary functions for display only ###

def cls_name(obj_or_cls):
    cls = type(obj_or_cls)
    if cls is type:
        cls = obj_or_cls
    return cls.__name__.split('.')[-1]

def display(obj):
    cls = type(obj)
    if cls is type:
        return f'<class {obj.__name__}>'
    elif cls in [type(None), int]:
        return repr(obj)
    else:
        return f'<{cls_name(obj)} object>'

def print_args(name, *args):
    pseudo_args = ', '.join(display(x) for x in args)
    print(f'-> {cls_name(args[0])}.__{name}__({pseudo_args})')


### essential classes for this example ###
```

```
class Overriding:  ❶
    """a.k.a. data descriptor or enforced descriptor"""

    def __get__(self, instance, owner):
        print_args('get', self, instance, owner)  ❷

    def __set__(self, instance, value):
        print_args('set', self, instance, value)


class OverridingNoGet:  ❸
    """an overriding descriptor without ``__get__``"""

    def __set__(self, instance, value):
        print_args('set', self, instance, value)


class NonOverriding:  ❹
    """a.k.a. non-data or shadowable descriptor"""

    def __get__(self, instance, owner):
        print_args('get', self, instance, owner)


class Managed:  ❺
    over = Overriding()
    over_no_get = OverridingNoGet()
    non_over = NonOverriding()

    def spam(self):  ❻
        print(f'-> Managed.spam({display(self)})')
```

❶ An overriding descriptor class with __get__ and __set__.

❷ The print_args function is called by every descriptor method in this example.

❸ An overriding descriptor without a __get__ method.

❹ No __set__ method here, so this is a nonoverriding descriptor.

❺ The managed class, using one instance of each of the descriptor classes.

❻ The spam method is here for comparison, because methods are also descriptors.

In the following sections, we will examine the behavior of attribute reads and writes on the Managed class, and one instance of it, going through each of the different descriptors defined.

Overriding Descriptors

A descriptor that implements the __set__ method is an *overriding descriptor*, because although it is a class attribute, a descriptor implementing __set__ will override attempts to assign to instance attributes. This is how Example 23-3 was implemented. Properties are also overriding descriptors: if you don't provide a setter function, the default __set__ from the property class will raise AttributeError to signal that the attribute is read-only. Given the code in Example 23-8, experiments with an overriding descriptor can be seen in Example 23-9.

 Python contributors and authors use different terms when discussing these concepts. I adopted "overriding descriptor" from the book *Python in a Nutshell*. The official Python documentation uses "data descriptor," but "overriding descriptor" highlights the special behavior. Overriding descriptors are also called "enforced descriptors." Synonyms for nonoverriding descriptors include "nondata descriptors" or "shadowable descriptors."

Example 23-9. Behavior of an overriding descriptor

```
>>> obj = Managed()   ❶
>>> obj.over   ❷
-> Overriding.__get__(<Overriding object>, <Managed object>, <class Managed>)
>>> Managed.over   ❸
-> Overriding.__get__(<Overriding object>, None, <class Managed>)
>>> obj.over = 7   ❹
-> Overriding.__set__(<Overriding object>, <Managed object>, 7)
>>> obj.over   ❺
-> Overriding.__get__(<Overriding object>, <Managed object>, <class Managed>)
>>> obj.__dict__['over'] = 8   ❻
>>> vars(obj)   ❼
{'over': 8}
>>> obj.over   ❽
-> Overriding.__get__(<Overriding object>, <Managed object>, <class Managed>)
```

❶ Create Managed object for testing.

❷ obj.over triggers the descriptor __get__ method, passing the managed instance obj as the second argument.

❸ Managed.over triggers the descriptor __get__ method, passing None as the second argument (instance).

❹ Assigning to obj.over triggers the descriptor __set__ method, passing the value 7 as the last argument.

❺ Reading `obj.over` still invokes the descriptor `__get__` method.

❻ Bypassing the descriptor, setting a value directly to the `obj.__dict__`.

❼ Verify that the value is in the `obj.__dict__`, under the `over` key.

❽ However, even with an instance attribute named `over`, the `Managed.over` descriptor still overrides attempts to read `obj.over`.

Overriding Descriptor Without __get__

Properties and other overriding descriptors, such as Django model fields, implement both `__set__` and `__get__`, but it's also possible to implement only `__set__`, as we saw in Example 23-2. In this case, only writing is handled by the descriptor. Reading the descriptor through an instance will return the descriptor object itself because there is no `__get__` to handle that access. If a namesake instance attribute is created with a new value via direct access to the instance `__dict__`, the `__set__` method will still override further attempts to set that attribute, but reading that attribute will simply return the new value from the instance, instead of returning the descriptor object. In other words, the instance attribute will shadow the descriptor, but only when reading. See Example 23-10.

Example 23-10. Overriding descriptor without __get__

```
>>> obj.over_no_get  ❶
<__main__.OverridingNoGet object at 0x665bcc>
>>> Managed.over_no_get  ❷
<__main__.OverridingNoGet object at 0x665bcc>
>>> obj.over_no_get = 7  ❸
-> OverridingNoGet.__set__(<OverridingNoGet object>, <Managed object>, 7)
>>> obj.over_no_get  ❹
<__main__.OverridingNoGet object at 0x665bcc>
>>> obj.__dict__['over_no_get'] = 9  ❺
>>> obj.over_no_get  ❻
9
>>> obj.over_no_get = 7  ❼
-> OverridingNoGet.__set__(<OverridingNoGet object>, <Managed object>, 7)
>>> obj.over_no_get  ❽
9
```

❶ This overriding descriptor doesn't have a `__get__` method, so reading `obj.over_no_get` retrieves the descriptor instance from the class.

❷ The same thing happens if we retrieve the descriptor instance directly from the managed class.

❸ Trying to set a value to `obj.over_no_get` invokes the `__set__` descriptor method.

❹ Because our `__set__` doesn't make changes, reading `obj.over_no_get` again retrieves the descriptor instance from the managed class.

❺ Going through the instance `__dict__` to set an instance attribute named `over_no_get`.

❻ Now that `over_no_get` instance attribute shadows the descriptor, but only for reading.

❼ Trying to assign a value to `obj.over_no_get` still goes through the descriptor set.

❽ But for reading, that descriptor is shadowed as long as there is a namesake instance attribute.

Nonoverriding Descriptor

A descriptor that does not implement `__set__` is a nonoverriding descriptor. Setting an instance attribute with the same name will shadow the descriptor, rendering it ineffective for handling that attribute in that specific instance. Methods and `@functools.cached_property` are implemented as nonoverriding descriptors. Example 23-11 shows the operation of a nonoverriding descriptor.

Example 23-11. Behavior of a nonoverriding descriptor

```
>>> obj = Managed()
>>> obj.non_over  ❶
-> NonOverriding.__get__(<NonOverriding object>, <Managed object>, <class Managed>)
>>> obj.non_over = 7  ❷
>>> obj.non_over  ❸
7
>>> Managed.non_over  ❹
-> NonOverriding.__get__(<NonOverriding object>, None, <class Managed>)
>>> del obj.non_over  ❺
>>> obj.non_over  ❻
-> NonOverriding.__get__(<NonOverriding object>, <Managed object>, <class Managed>)
```

❶ `obj.non_over` triggers the descriptor `__get__` method, passing `obj` as the second argument.

❷ `Managed.non_over` is a nonoverriding descriptor, so there is no `__set__` to interfere with this assignment.

❸ The obj now has an instance attribute named non_over, which shadows the namesake descriptor attribute in the Managed class.

❹ The Managed.non_over descriptor is still there, and catches this access via the class.

❺ If the non_over instance attribute is deleted…

❻ …then reading obj.non_over hits the __get__ method of the descriptor in the class, but note that the second argument is the managed instance.

In the previous examples, we saw several assignments to an instance attribute with the same name as a descriptor, and different results according to the presence of a __set__ method in the descriptor.

The setting of attributes in the class cannot be controlled by descriptors attached to the same class. In particular, this means that the descriptor attributes themselves can be clobbered by assigning to the class, as the next section explains.

Overwriting a Descriptor in the Class

Regardless of whether a descriptor is overriding or not, it can be overwritten by assignment to the class. This is a monkey-patching technique, but in Example 23-12 the descriptors are replaced by integers, which would effectively break any class that depended on the descriptors for proper operation.

Example 23-12. Any descriptor can be overwritten on the class itself

```
>>> obj = Managed()   ❶
>>> Managed.over = 1   ❷
>>> Managed.over_no_get = 2
>>> Managed.non_over = 3
>>> obj.over, obj.over_no_get, obj.non_over   ❸
(1, 2, 3)
```

❶ Create a new instance for later testing.

❷ Overwrite the descriptor attributes in the class.

❸ The descriptors are really gone.

Example 23-12 reveals another asymmetry regarding reading and writing attributes: although the reading of a class attribute can be controlled by a descriptor with __get__ attached to the managed class, the writing of a class attribute cannot be handled by a descriptor with __set__ attached to the same class.

In order to control the setting of attributes in a class, you have to attach descriptors to the class of the class—in other words, the metaclass. By default, the metaclass of user-defined classes is type, and you cannot add attributes to type. But in Chapter 24, we'll create our own metaclasses.

Let's now focus on how descriptors are used to implement methods in Python.

Methods Are Descriptors

A function within a class becomes a bound method when invoked on an instance because all user-defined functions have a __get__ method, therefore they operate as descriptors when attached to a class. Example 23-13 demonstrates reading the spam method from the Managed class introduced in Example 23-8.

Example 23-13. A method is a nonoverriding descriptor

```
>>> obj = Managed()
>>> obj.spam    ❶
<bound method Managed.spam of <descriptorkinds.Managed object at 0x74c80c>>
>>> Managed.spam    ❷
<function Managed.spam at 0x734734>
>>> obj.spam = 7    ❸
>>> obj.spam
7
```

❶ Reading from obj.spam retrieves a bound method object.

❷ But reading from Managed.spam retrieves a function.

❸ Assigning a value to obj.spam shadows the class attribute, rendering the spam method inaccessible from the obj instance.

Functions do not implement __set__, therefore they are nonoverriding descriptors, as the last line of Example 23-13 shows.

The other key takeaway from Example 23-13 is that obj.spam and Managed.spam retrieve different objects. As usual with descriptors, the __get__ of a function returns a reference to itself when the access happens through the managed class. But when the access goes through an instance, the __get__ of the function returns a bound method object: a callable that wraps the function and binds the managed instance (e.g., obj) to the first argument of the function (i.e., self), like the functools.partial function does (as seen in "Freezing Arguments with functools.partial" on page 247). For a deeper understanding of this mechanism, take a look at Example 23-14.

Example 23-14. method_is_descriptor.py: a Text class, derived from UserString

```
import collections


class Text(collections.UserString):

    def __repr__(self):
        return 'Text({!r})'.format(self.data)

    def reverse(self):
        return self[::-1]
```

Now let's investigate the Text.reverse method. See Example 23-15.

Example 23-15. Experiments with a method

```
>>> word = Text('forward')
>>> word  ❶
Text('forward')
>>> word.reverse()  ❷
Text('drawrof')
>>> Text.reverse(Text('backward'))  ❸
Text('drawkcab')
>>> type(Text.reverse), type(word.reverse)  ❹
(<class 'function'>, <class 'method'>)
>>> list(map(Text.reverse, ['repaid', (10, 20, 30), Text('stressed')]))  ❺
['diaper', (30, 20, 10), Text('desserts')]
>>> Text.reverse.__get__(word)  ❻
<bound method Text.reverse of Text('forward')>
>>> Text.reverse.__get__(None, Text)  ❼
<function Text.reverse at 0x101244e18>
>>> word.reverse  ❽
<bound method Text.reverse of Text('forward')>
>>> word.reverse.__self__  ❾
Text('forward')
>>> word.reverse.__func__ is Text.reverse  ❿
True
```

❶ The repr of a Text instance looks like a Text constructor call that would make an equal instance.

❷ The reverse method returns the text spelled backward.

❸ A method called on the class works as a function.

❹ Note the different types: a function and a method.

❺ `Text.reverse` operates as a function, even working with objects that are not instances of `Text`.

❻ Any function is a nonoverriding descriptor. Calling its `__get__` with an instance retrieves a method bound to that instance.

❼ Calling the function's `__get__` with `None` as the `instance` argument retrieves the function itself.

❽ The expression `word.reverse` actually invokes `Text.reverse.__get__(word)`, returning the bound method.

❾ The bound method object has a `__self__` attribute holding a reference to the instance on which the method was called.

❿ The `__func__` attribute of the bound method is a reference to the original function attached to the managed class.

The bound method object also has a `__call__` method, which handles the actual invocation. This method calls the original function referenced in `__func__`, passing the `__self__` attribute of the method as the first argument. That's how the implicit binding of the conventional `self` argument works.

The way functions are turned into bound methods is a prime example of how descriptors are used as infrastructure in the language.

After this deep dive into how descriptors and methods work, let's go through some practical advice about their use.

Descriptor Usage Tips

The following list addresses some practical consequences of the descriptor characteristics just described:

Use `property` to keep it simple
The `property` built-in creates overriding descriptors implementing `__set__` and `__get__` even if you do not define a setter method.[7] The default `__set__` of a property raises `AttributeError: can't set attribute`, so a property is the easiest way to create a read-only attribute, avoiding the issue described next.

---

7 A `__delete__` method is also provided by the `property` decorator, even if no deleter method is defined by you.

Read-only descriptors require __set__

If you use a descriptor class to implement a read-only attribute, you must remember to code both __get__ and __set__, otherwise setting a namesake attribute on an instance will shadow the descriptor. The __set__ method of a read-only attribute should just raise AttributeError with a suitable message.[8]

Validation descriptors can work with __set__ only

In a descriptor designed only for validation, the __set__ method should check the value argument it gets, and if valid, set it directly in the instance __dict__ using the descriptor instance name as key. That way, reading the attribute with the same name from the instance will be as fast as possible, because it will not require a __get__. See the code for Example 23-3.

Caching can be done efficiently with __get__ only

If you code just the __get__ method, you have a nonoverriding descriptor. These are useful to make some expensive computation and then cache the result by setting an attribute by the same name on the instance.[9] The namesake instance attribute will shadow the descriptor, so subsequent access to that attribute will fetch it directly from the instance __dict__ and not trigger the descriptor __get__ anymore. The @functools.cached_property decorator actually produces a nonoverriding descriptor.

Nonspecial methods can be shadowed by instance attributes

Because functions and methods only implement __get__, they are nonoverriding descriptors. A simple assignment like my_obj.the_method = 7 means that further access to the_method through that instance will retrieve the number 7—without affecting the class or other instances. However, this issue does not interfere with special methods. The interpreter only looks for special methods in the class itself, in other words, repr(x) is executed as x.__class__.__repr__(x), so a __repr__ attribute defined in x has no effect on repr(x). For the same reason, the existence of an attribute named __getattr__ in an instance will not subvert the usual attribute access algorithm.

The fact that nonspecial methods can be overridden so easily in instances may sound fragile and error prone, but I personally have never been bitten by this in more than 20 years of Python coding. On the other hand, if you are doing a lot of dynamic

---

8 Python is not consistent in such messages. Trying to change the c.real attribute of a complex number gets AttributeError: readonly attribute, but an attempt to change c.conjugate (a method of complex), results in AttributeError: 'complex' object attribute 'conjugate' is read-only. Even the spelling of "read-only" is different.

9 However, recall that creating instance attributes after the __init__ method runs defeats the key-sharing memory optimization, as discussed in from "Practical Consequences of How dict Works" on page 102.

attribute creation, where the attribute names come from data you don't control (as we did in the earlier parts of this chapter), then you should be aware of this and perhaps implement some filtering or escaping of the dynamic attribute names to preserve your sanity.

 The FrozenJSON class in Example 22-5 is safe from instance attribute shadowing methods because its only methods are special methods and the build class method. Class methods are safe as long as they are always accessed through the class, as I did with FrozenJSON.build in Example 22-5—later replaced by __new__ in Example 22-6. The Record and Event classes presented in "Computed Properties" on page 849 are also safe: they implement only special methods, static methods, and properties. Properties are overriding descriptors, so they are not shadowed by instance attributes.

To close this chapter, we'll cover two features we saw with properties that we have not addressed in the context of descriptors: documentation and handling attempts to delete a managed attribute.

Descriptor Docstring and Overriding Deletion

The docstring of a descriptor class is used to document every instance of the descriptor in the managed class. Figure 23-4 shows the help displays for the LineItem class with the Quantity and NonBlank descriptors from Examples 23-6 and 23-7.

That is somewhat unsatisfactory. In the case of LineItem, it would be good to add, for example, the information that weight must be in kilograms. That would be trivial with properties, because each property handles a specific managed attribute. But with descriptors, the same Quantity descriptor class is used for weight and price.[10]

The second detail we discussed with properties, but have not addressed with descriptors, is handling attempts to delete a managed attribute. That can be done by implementing a __delete__ method alongside or instead of the usual __get__ and/or __set__ in the descriptor class. I deliberately omitted coverage of __delete__ because I believe real-world usage·is rare. If you need this, please see the "Implementing Descriptors" (*https://fpy.li/23-2*) section of the Python Data Model documentation (*https://fpy.li/dtmodel*). Coding a silly descriptor class with __delete__ is left as an exercise to the leisurely reader.

---

10 Customizing the help text for each descriptor instance is surprisingly hard. One solution requires dynamically building a wrapper class for each descriptor instance.

Figure 23-4. Screenshots of the Python console when issuing the commands `help(LineI` `tem.weight)` *and* `help(LineItem)`.

Chapter Summary

The first example of this chapter was a continuation of the `LineItem` examples from Chapter 22. In Example 23-2, we replaced properties with descriptors. We saw that a descriptor is a class that provides instances that are deployed as attributes in the managed class. Discussing this mechanism required special terminology, introducing terms such as *managed instance* and *storage attribute*.

In "LineItem Take #4: Automatic Naming of Storage Attributes" on page 891, we removed the requirement that `Quantity` descriptors were declared with an explicit `storage_name`, which was redundant and error prone. The solution was to implement the `__set_name__` special method in `Quantity`, to save the name of the managed property as `self.storage_name`.

"LineItem Take #5: A New Descriptor Type" on page 893 showed how to subclass an abstract descriptor class to share code while building specialized descriptors with some common functionality.

We then looked at the different behaviors of descriptors providing or omitting the __set__ method, making the crucial distinction between overriding and nonoverriding descriptors, a.k.a. data and nondata descriptors. Through detailed testing we uncovered when descriptors are in control and when they are shadowed, bypassed, or overwritten.

Following that, we studied a particular category of nonoverriding descriptors: methods. Console experiments revealed how a function attached to a class becomes a method when accessed through an instance, by leveraging the descriptor protocol.

To conclude the chapter, "Descriptor Usage Tips" on page 904 presented practical tips, and "Descriptor Docstring and Overriding Deletion" on page 906 provided a brief look at how to document descriptors.

As noted in "What's New in This Chapter" on page 884, several examples in this chapter became much simpler thanks to the __set_name__ special method of the descriptor protocol, added in Python 3.6. That's language evolution!

Further Reading

Besides the obligatory reference to the "Data Model" chapter (*https://fpy.li/dtmodel*), Raymond Hettinger's "Descriptor HowTo Guide" (*https://fpy.li/23-3*) is a valuable resource—part of the HowTo collection (*https://fpy.li/23-4*) in the official Python documentation.

As usual with Python object model subjects, Martelli, Ravenscroft, and Holden's *Python in a Nutshell*, 3rd ed. (O'Reilly) is authoritative and objective. Martelli also has a presentation titled "Python's Object Model," which covers properties and descriptors in depth (see the slides (*https://fpy.li/23-5*) and video (*https://fpy.li/23-6*)).

Beware that any coverage of descriptors written or recorded before PEP 487 was adopted in 2016 is likely to contain examples that are needlessly complicated today, because __set_name__ was not supported in Python versions prior to 3.6.

For more practical examples, *Python Cookbook*, 3rd ed., by David Beazley and Brian K. Jones (O'Reilly), has many recipes illustrating descriptors, of which I want to highlight "6.12. Reading Nested and Variable-Sized Binary Structures," "8.10. Using Lazily Computed Properties," "8.13. Implementing a Data Model or Type System," and "9.9. Defining Decorators As Classes." The last recipe of which addresses deep issues with the interaction of function decorators, descriptors, and methods, explaining how a function decorator implemented as a class with __call__ also needs to implement __get__ if it wants to work with decorating methods as well as functions.

PEP 487—Simpler customization of class creation (*https://fpy.li/pep487*) introduced the __set_name__ special method, and includes an example of a validating descriptor (*https://fpy.li/23-7*).

Soapbox

The Design of self

The requirement to explicitly declare self as a first argument in methods is a controversial design decision in Python. After 23 years using the language, I am used to it. I think that decision is an example of "worse is better": a design philosophy described by computer scientist Richard P. Gabriel in "The Rise of Worse is Better" (*https://fpy.li/23-8*). The first priority of this philosophy is "simplicity," which Gabriel presents as:

> The design must be simple, both in implementation and interface. It is more important for the implementation to be simple than the interface. Simplicity is the most important consideration in a design.

Python's explicit self embodies that design philosophy. The implementation is simple—elegant even—at the expense of the user interface: a method signature like def zfill(self, width): doesn't visually match the invocation label.zfill(8).

Modula-3 introduced that convention with the same identifier self. But there is a key difference: in Modula-3, interfaces are declared separately from their implementation, and in the interface declaration the self argument is omitted, so from the user's perspective, a method appears in an interface declaration with the same explicit parameters used to call it.

Over time, Python's error messages related to method arguments became clearer. For a user-defined method with one argument besides self, if the user invokes obj.meth(), Python 2.7 raised:

```
TypeError: meth() takes exactly 2 arguments (1 given)
```

In Python 3, the confusing argument count is not mentioned, and the missing argument is named:

```
TypeError: meth() missing 1 required positional argument: 'x'
```

Besides the use of self as an explicit argument, the requirement to qualify every access to instance attributes with self is also criticized. See, for example, A. M. Kuchling's famous "Python Warts" post (archived (*https://fpy.li/23-9*)); Kuchling himself is not so bothered by the self qualifier, but he mentions it—probably echoing opinions from the comp.lang.python group. I personally don't mind typing the self qualifier:

it's good to distinguish local variables from attributes. My issue is with the use of self in the def statement.

Anyone who is unhappy about the explicit self in Python can feel a lot better by considering the baffling semantics (*https://fpy.li/23-10*) of the implicit this in JavaScript. Guido had some good reasons to make self work as it does, and he wrote about them in "Adding Support for User-Defined Classes" (*https://fpy.li/23-11*), a post on his blog, *The History of Python*.

## CHAPTER 24

Class Metaprogramming

Everyone knows that debugging is twice as hard as writing a program in the first place. So if you're as clever as you can be when you write it, how will you ever debug it?

—Brian W. Kernighan and P. J. Plauger, *The Elements of Programming Style*[1]

Class metaprogramming is the art of creating or customizing classes at runtime. Classes are first-class objects in Python, so a function can be used to create a new class at any time, without using the `class` keyword. Class decorators are also functions, but designed to inspect, change, and even replace the decorated class with another class. Finally, metaclasses are the most advanced tool for class metaprogramming: they let you create whole new categories of classes with special traits, such as the abstract base classes we've already seen.

Metaclasses are powerful, but hard to justify and even harder to get right. Class decorators solve many of the same problems and are easier to understand. Furthermore, Python 3.6 implemented PEP 487—Simpler customization of class creation (*https://fpy.li/pep487*), providing special methods supporting tasks that previously required metaclasses or class decorators.[2]

This chapter presents the class metaprogramming techniques in ascending order of complexity.

---

1 Quote from Chapter 2, "Expression" of *The Elements of Programming Style*, 2nd ed. (McGraw-Hill), page 10.

2 That doesn't mean PEP 487 broke code that used those features. It just means that some code that used class decorators or metaclasses prior to Python 3.6 can now be refactored to use plain classes, resulting in simpler and possibly more efficient code.

This is an exciting topic, and it's easy to get carried away. So I must offer this advice.

For the sake of readability and maintainability, you should probably avoid the techniques described in this chapter in application code.

On the other hand, these are the tools of the trade if you want to write the next great Python framework.

What's New in This Chapter

All the code in the "Class Metaprogramming" chapter of the first edition of *Fluent Python* still runs correctly. However, some of the previous examples no longer represent the simplest solutions in light of new features added since Python 3.6.

I replaced those examples with different ones, highlighting Python's new metaprogramming features or adding further requirements to justify the use of the more advanced techniques. Some of the new examples leverage type hints to provide class builders similar to the @dataclass decorator and typing.NamedTuple.

"Metaclasses in the Real World" on page 951 is a new section with some high-level considerations about the applicability of metaclasses.

Some of the best refactorings are removing code made redundant by newer and simpler ways of solving the same problems. This applies to production code as well as books.

We'll get started by reviewing attributes and methods defined in the Python Data Model for all classes.

Classes as Objects

Like most program entities in Python, classes are also objects. Every class has a number of attributes defined in the Python Data Model, documented in "4.13. Special Attributes" (*https://fpy.li/24-1*) of the "Built-in Types" chapter in *The Python Standard Library*. Three of those attributes appeared several times in this book already: __class__, __name__, and __mro__. Other class standard attributes are:

cls.__bases__
    The tuple of base classes of the class.

`cls.__qualname__`
> The qualified name of a class or function, which is a dotted path from the global scope of the module to the class definition. This is relevant when the class is defined inside another class. For example, in a Django model class such as Ox (*https://fpy.li/24-2*), there is an inner class called Meta. The `__qualname__` of Meta is Ox.Meta, but its `__name__` is just Meta. The specification for this attribute is PEP 3155—Qualified name for classes and functions (*https://fpy.li/24-3*).

`cls.__subclasses__()`
> This method returns a list of the immediate subclasses of the class. The implementation uses weak references to avoid circular references between the superclass and its subclasses—which hold a strong reference to the superclasses in their `__bases__` attribute. The method lists subclasses currently in memory. Subclasses in modules not yet imported will not appear in the result.

`cls.mro()`
> The interpreter calls this method when building a class to obtain the tuple of superclasses stored in the `__mro__` attribute of the class. A metaclass can override this method to customize the method resolution order of the class under construction.

> None of the attributes mentioned in this section are listed by the `dir(…)` function.

Now, if a class is an object, what is the class of a class?

type: The Built-In Class Factory

We usually think of `type` as a function that returns the class of an object, because that's what `type(my_object)` does: it returns `my_object.__class__`.

However, `type` is a class that creates a new class when invoked with three arguments.

Consider this simple class:

```
class MyClass(MySuperClass, MyMixin):
    x = 42

    def x2(self):
        return self.x * 2
```

Using the `type` constructor, you can create `MyClass` at runtime with this code:

```
MyClass = type('MyClass',
               (MySuperClass, MyMixin),
               {'x': 42, 'x2': lambda self: self.x * 2},
              )
```

That `type` call is functionally equivalent to the previous `class MyClass…` block statement.

When Python reads a `class` statement, it calls `type` to build the class object with these parameters:

`name`
> The identifier that appears after the `class` keyword, e.g., `MyClass`.

`bases`
> The tuple of superclasses given in parentheses after the class identifier, or `(object,)` if superclasses are not mentioned in the `class` statement.

`dict`
> A mapping of attribute names to values. Callables become methods, as we saw in "Methods Are Descriptors" on page 902. Other values become class attributes.

 The `type` constructor accepts optional keyword arguments, which are ignored by `type` itself, but are passed untouched into `__init_subclass__`, which must consume them. We'll study that special method in "Introducing `__init_subclass__`" on page 918, but I won't cover the use of keyword arguments. For more, please read PEP 487—Simpler customization of class creation (*https://fpy.li/ pep487*).

The `type` class is a *metaclass*: a class that builds classes. In other words, instances of the `type` class are classes. The standard library provides a few other metaclasses, but `type` is the default:

```
>>> type(7)
<class 'int'>
>>> type(int)
<class 'type'>
>>> type(OSError)
<class 'type'>
>>> class Whatever:
...     pass
...
>>> type(Whatever)
<class 'type'>
```

We'll build custom metaclasses in "Metaclasses 101" on page 935.

Next, we'll use the `type` built-in to make a function that builds classes.

A Class Factory Function

The standard library has a class factory function that appears several times in this book: collections.namedtuple. In Chapter 5 we also saw typing.NamedTuple and @dataclass. All of these class builders leverage techniques covered in this chapter.

We'll start with a super simple factory for classes of mutable objects—the simplest possible replacement for @dataclass.

Suppose I'm writing a pet shop application and I want to store data for dogs as simple records. But I don't want to write boilerplate like this:

```
class Dog:
    def __init__(self, name, weight, owner):
        self.name = name
        self.weight = weight
        self.owner = owner
```

Boring...each field name appears three times, and that boilerplate doesn't even buy us a nice repr:

```
>>> rex = Dog('Rex', 30, 'Bob')
>>> rex
<__main__.Dog object at 0x2865bac>
```

Taking a hint from collections.namedtuple, let's create a record_factory that creates simple classes like Dog on the fly. Example 24-1 shows how it should work.

Example 24-1. Testing record_factory, a simple class factory

```
>>> Dog = record_factory('Dog', 'name weight owner')   ❶
>>> rex = Dog('Rex', 30, 'Bob')
>>> rex   ❷
Dog(name='Rex', weight=30, owner='Bob')
>>> name, weight, _ = rex   ❸
>>> name, weight
('Rex', 30)
>>> "{2}'s dog weighs {1}kg".format(*rex)   ❹
"Bob's dog weighs 30kg"
>>> rex.weight = 32   ❺
>>> rex
Dog(name='Rex', weight=32, owner='Bob')
>>> Dog.__mro__   ❻
(<class 'factories.Dog'>, <class 'object'>)
```

❶ Factory can be called like namedtuple: class name, followed by attribute names separated by spaces in a single string.

❷ Nice repr.

❸ Instances are iterable, so they can be conveniently unpacked on assignment…

❹ …or when passing to functions like format.

❺ A record instance is mutable.

❻ The newly created class inherits from object—no relationship to our factory.

The code for record_factory is in Example 24-2.[3]

Example 24-2. record_factory.py: a simple class factory

```python
from typing import Union, Any
from collections.abc import Iterable, Iterator

FieldNames = Union[str, Iterable[str]]  ❶

def record_factory(cls_name: str, field_names: FieldNames) -> type[tuple]:  ❷

    slots = parse_identifiers(field_names)  ❸

    def __init__(self, *args, **kwargs) -> None:  ❹
        attrs = dict(zip(self.__slots__, args))
        attrs.update(kwargs)
        for name, value in attrs.items():
            setattr(self, name, value)

    def __iter__(self) -> Iterator[Any]:  ❺
        for name in self.__slots__:
            yield getattr(self, name)

    def __repr__(self):  ❻
        values = ', '.join(f'{name}={value!r}'
            for name, value in zip(self.__slots__, self))
        cls_name = self.__class__.__name__
        return f'{cls_name}({values})'

    cls_attrs = dict(  ❼
        __slots__=slots,
        __init__=__init__,
        __iter__=__iter__,
        __repr__=__repr__,
    )

    return type(cls_name, (object,), cls_attrs)  ❽
```

---

3 Thanks to my friend J. S. O. Bueno for contributing to this example.

```
def parse_identifiers(names: FieldNames) -> tuple[str, ...]:
    if isinstance(names, str):
        names = names.replace(',', ' ').split()  ❾
    if not all(s.isidentifier() for s in names):
        raise ValueError('names must all be valid identifiers')
    return tuple(names)
```

❶ User can provide field names as a single string or an iterable of strings.

❷ Accept arguments like the first two of collections.namedtuple; return a type—
    i.e., a class that behaves like a tuple.

❸ Build a tuple of attribute names; this will be the __slots__ attribute of the new
    class.

❹ This function will become the __init__ method in the new class. It accepts posi-
    tional and/or keyword arguments.[4]

❺ Yield the field values in the order given by __slots__.

❻ Produce the nice repr, iterating over __slots__ and self.

❼ Assemble a dictionary of class attributes.

❽ Build and return the new class, calling the type constructor.

❾ Convert names separated by spaces or commas to list of str.

Example 24-2 is the first time we've seen type in a type hint. If the annotation was
just -> type, that would mean that record_factory returns a class—and it would be
correct. But the annotation -> type[tuple] is more precise: it says the returned class
will be a subclass of tuple.

The last line of record_factory in Example 24-2 builds a class named by the value of
cls_name, with object as its single immediate base class, and with a namespace
loaded with __slots__, __init__, __iter__, and __repr__, of which the last three
are instance methods.

We could have named the __slots__ class attribute anything else, but then we'd have
to implement __setattr__ to validate the names of attributes being assigned,

---

4 I did not add type hints to the arguments because the actual types are Any. I put the return type hint because
  otherwise Mypy will not check inside the method.

because for our record-like classes we want the set of attributes to be always the same and in the same order. However, recall that the main feature of __slots__ is saving memory when you are dealing with millions of instances, and using __slots__ has some drawbacks, discussed in "Saving Memory with __slots__" on page 386.

 Instances of classes created by record_factory are not serializable —that is, they can't be exported with the dump function from the pickle module. Solving this problem is beyond the scope of this example, which aims to show the type class in action in a simple use case. For the full solution, study the source code for collections.namedtuple (*https://fpy.li/24-4*); search for the word "pickling."

Now let's see how to emulate more modern class builders like typing.NamedTuple, which takes a user-defined class written as a class statement, and automatically enhances it with more functionality.

Introducing __init_subclass__

Both __init_subclass__ and __set_name__ were proposed in PEP 487—Simpler customization of class creation (*https://fpy.li/pep487*). We saw the __set_name__ special method for descriptors for the first time in "LineItem Take #4: Automatic Naming of Storage Attributes" on page 891. Now let's study __init_subclass__.

In Chapter 5, we saw that typing.NamedTuple and @dataclass let programmers use the class statement to specify attributes for a new class, which is then enhanced by the class builder with the automatic addition of essential methods like __init__, __repr__, __eq__, etc.

Both of these class builders read type hints in the user's class statement to enhance the class. Those type hints also allow static type checkers to validate code that sets or gets those attributes. However, NamedTuple and @dataclass do not take advantage of the type hints for attribute validation at runtime. The Checked class in the next example does.

 It is not possible to support every conceivable static type hint for runtime type checking, which is probably why typing.NamedTuple and @dataclass don't even try it. However, some types that are also concrete classes can be used with Checked. This includes simple types often used for field contents, such as str, int, float, and bool, as well as lists of those types.

Example 24-3 shows how to use Checked to build a Movie class.

Example 24-3. initsub/checkedlib.py: doctest for creating a Movie subclass of Checked

```
>>> class Movie(Checked):      ❶
...     title: str      ❷
...     year: int
...     box_office: float
...
>>> movie = Movie(title='The Godfather', year=1972, box_office=137)      ❸
>>> movie.title
'The Godfather'
>>> movie      ❹
Movie(title='The Godfather', year=1972, box_office=137.0)
```

❶ Movie inherits from Checked—which we'll define later in Example 24-5.

❷ Each attribute is annotated with a constructor. Here I used built-in types.

❸ Movie instances must be created using keyword arguments.

❹ In return, you get a nice __repr__.

The constructors used as the attribute type hints may be any callable that takes zero or one argument and returns a value suitable for the intended field type, or rejects the argument by raising TypeError or ValueError.

Using built-in types for the annotations in Example 24-3 means the values must be acceptable by the constructor of the type. For int, this means any x such that int(x) returns an int. For str, anything goes at runtime, because str(x) works with any x in Python.[5]

When called with no arguments, the constructor should return a default value of its type.[6]

This is standard behavior for Python's built-in constructors:

```
>>> int(), float(), bool(), str(), list(), dict(), set()
(0, 0.0, False, '', [], {}, set())
```

---

5 That's true for any object, except when its class overrides the __str__ or __repr__ methods inherited from object with broken implementations.

6 This solution avoids using None as a default. Avoiding null values is a good idea (*https://fpy.li/24-5*). They are hard to avoid in general, but easy in some cases. In Python as well as SQL, I prefer to represent missing data in a text field with an empty string instead of None or NULL. Learning Go reinforced this idea: variables and struct fields of primitive types in Go are initialized by default with a "zero value." See "Zero values" in the online *Tour of Go* (*https://fpy.li/24-6*) if you are curious.

In a Checked subclass like Movie, missing parameters create instances with default values returned by the field constructors. For example:

```
>>> Movie(title='Life of Brian')
Movie(title='Life of Brian', year=0, box_office=0.0)
```

The constructors are used for validation during instantiation and when an attribute is set directly on an instance:

```
>>> blockbuster = Movie(title='Avatar', year=2009, box_office='billions')
Traceback (most recent call last):
  ...
TypeError: 'billions' is not compatible with box_office:float
>>> movie.year = 'MCMLXXII'
Traceback (most recent call last):
  ...
TypeError: 'MCMLXXII' is not compatible with year:int
```

Checked Subclasses and Static Type Checking

In a *.py* source file with a movie instance of Movie, as defined in Example 24-3, Mypy flags this assignment as a type error:

```
movie.year = 'MCMLXXII'
```

However, Mypy can't detect type errors in this constructor call:

```
blockbuster = Movie(title='Avatar', year='MMIX')
```

That's because Movie inherits Checked.__init__, and the signature of that method must accept any keyword arguments to support arbitrary user-defined classes.

On the other hand, if you declare a Checked subclass field with the type hint list[float], Mypy can flag assignments of lists with incompatible contents, but Checked will ignore the type parameter and treat that the same as list.

Now let's look at the implementation of *checkedlib.py*. The first class is the Field descriptor, as shown in Example 24-4.

Example 24-4. initsub/checkedlib.py: the Field descriptor class

```
from collections.abc import Callable  ❶
from typing import Any, NoReturn, get_type_hints


class Field:
    def __init__(self, name: str, constructor: Callable) -> None:  ❷
        if not callable(constructor) or constructor is type(None):  ❸
            raise TypeError(f'{name!r} type hint must be callable')
        self.name = name
```

```
        self.constructor = constructor

    def __set__(self, instance: Any, value: Any) -> None:
        if value is ...:        ❹
            value = self.constructor()
        else:
            try:
                value = self.constructor(value)        ❺
            except (TypeError, ValueError) as e:        ❻
                type_name = self.constructor.__name__
                msg = f'{value!r} is not compatible with {self.name}:{type_name}'
                raise TypeError(msg) from e
        instance.__dict__[self.name] = value        ❼
```

❶ Recall that since Python 3.9, the Callable type for annotations is the ABC in collections.abc, and not the deprecated typing.Callable.

❷ This is a minimal Callable type hint; the parameter type and return type for constructor are both implicitly Any.

❸ For runtime checking, we use the callable built-in.[7] The test against type(None) is necessary because Python reads None in a type as NoneType, the class of None (therefore callable), but a useless constructor that only returns None.

❹ If Checked.__init__ sets the value as ... (the Ellipsis built-in object), we call the constructor with no arguments.

❺ Otherwise, call the constructor with the given value.

❻ If constructor raises either of these exceptions, we raise TypeError with a helpful message including the names of the field and constructor; e.g., 'MMIX' is not compatible with year:int.

❼ If no exceptions were raised, the value is stored in the instance.__dict__.

In __set__, we need to catch TypeError and ValueError because built-in constructors may raise either of them, depending on the argument. For example, float(None) raises TypeError, but float('A') raises ValueError. On the other hand, float('8') raises no error and returns 8.0. I hereby declare that this is a feature and not a bug of this toy example.

---

7 I believe that callable should be made suitable for type hinting. As of May 6, 2021, this is an open issue (*https://fpy.li/24-7*).

In "LineItem Take #4: Automatic Naming of Storage Attributes" on page 891, we saw the handy __set_name__ special method for descriptors. We don't need it in the Field class because the descriptors are not instantiated in client source code; the user declares types that are constructors, as we saw in the Movie class (Example 24-3). Instead, the Field descriptor instances are created at runtime by the Checked.__init_subclass__ method, which we'll see in Example 24-5.

Now let's focus on the Checked class. I split it in two listings. Example 24-5 shows the top of the class, which includes the most important methods in this example. The remaining methods are in Example 24-6.

Example 24-5. initsub/checkedlib.py: the most important methods of the Checked class

```python
class Checked:
    @classmethod
    def _fields(cls) -> dict[str, type]:      ❶
        return get_type_hints(cls)

    def __init_subclass__(subclass) -> None:   ❷
        super().__init_subclass__()            ❸
        for name, constructor in subclass._fields().items():   ❹
            setattr(subclass, name, Field(name, constructor))  ❺

    def __init__(self, **kwargs: Any) -> None:
        for name in self._fields():            ❻
            value = kwargs.pop(name, ...)       ❼
            setattr(self, name, value)          ❽
        if kwargs:                              ❾
            self.__flag_unknown_attrs(*kwargs)  ❿
```

❶ I wrote this class method to hide the call to typing.get_type_hints from the rest of the class. If I need to support Python ≥ 3.10 only, I'd call inspect.get_annotations instead. Review "Problems with Annotations at Runtime" on page 542 for the issues with those functions.

❷ __init_subclass__ is called when a subclass of the current class is defined. It gets that new subclass as its first argument—which is why I named the argument subclass instead of the usual cls. For more on this, see "__init_subclass__ Is Not a Typical Class Method" on page 923.

❸ super().__init_subclass__() is not strictly necessary, but should be invoked to play nice with other classes that might implement .__init_subclass__() in the same inheritance graph. See "Multiple Inheritance and Method Resolution Order" on page 496.

❹ Iterate over each field `name` and `constructor`…

❺ …creating an attribute on `subclass` with that `name` bound to a `Field` descriptor parameterized with `name` and `constructor`.

❻ For each `name` in the class fields…

❼ …get the corresponding `value` from `kwargs` and remove it from `kwargs`. Using `...` (the `Ellipsis` object) as default allows us to distinguish between arguments given the value `None` from arguments that were not given.[8]

❽ This `setattr` call triggers `Checked.__setattr__`, shown in Example 24-6.

❾ If there are remaining items in `kwargs`, their names do not match any of the declared fields, and `__init__` will fail.

❿ The error is reported by `__flag_unknown_attrs`, listed in Example 24-6. It takes a `*names` argument with the unknown attribute names. I used a single asterisk in `*kwargs` to pass its keys as a sequence of arguments.

`__init_subclass__` Is Not a Typical Class Method

The `@classmethod` decorator is never used with `__init_subclass__`, but that doesn't mean much, because the `__new__` special method behaves as a class method even without `@classmethod`. The first argument that Python passes to `__init_subclass__` is a class. However, it is never the class where `__init_subclass__` is implemented: it is a newly defined subclass of that class. That's unlike `__new__` and every other class method that I know about. Therefore, I think `__init_subclass__` is not a class method in the usual sense, and it is misleading to name the first argument `cls`. The `__init_suclass__` documentation (*https://fpy.li/24-8*) names the argument `cls` but explains: "…called whenever the containing class is subclassed. cls is then the new subclass."

---

8 As mentioned in "Loops, Sentinels, and Poison Pills" on page 725, the `Ellipsis` object is a convenient and safe sentinel value. It has been around for a long time, but recently people are finding more uses for it, as we see in type hints and NumPy.

Now let's see the remaining methods of the Checked class, continuing from Example 24-5. Note that I prepended _ to the _fields and _asdict method names for the same reason the collections.namedtuple API does: to reduce the chance of name clashes with user-defined field names.

Example 24-6. initsub/checkedlib.py: remaining methods of the Checked class

```
def __setattr__(self, name: str, value: Any) -> None:   ❶
    if name in self._fields():                          ❷
        cls = self.__class__
        descriptor = getattr(cls, name)
        descriptor.__set__(self, value)                 ❸
    else:                                               ❹
        self.__flag_unknown_attrs(name)

def __flag_unknown_attrs(self, *names: str) -> NoReturn:   ❺
    plural = 's' if len(names) > 1 else ''
    extra = ', '.join(f'{name!r}' for name in names)
    cls_name = repr(self.__class__.__name__)
    raise AttributeError(f'{cls_name} object has no attribute{plural} {extra}')

def _asdict(self) -> dict[str, Any]:   ❻
    return {
        name: getattr(self, name)
        for name, attr in self.__class__.__dict__.items()
        if isinstance(attr, Field)
    }

def __repr__(self) -> str:   ❼
    kwargs = ', '.join(
        f'{key}={value!r}' for key, value in self._asdict().items()
    )
    return f'{self.__class__.__name__}({kwargs})'
```

❶ Intercept all attempts to set an instance attribute. This is needed to prevent setting an unknown attribute.

❷ If the attribute name is known, fetch the corresponding descriptor.

❸ Usually we don't need to call the descriptor __set__ explicitly. It was necessary in this case because __setattr__ intercepts all attempts to set an attribute on the instance, including in the presence of an overriding descriptor such as Field.[9]

---

9 The subtle concept of an overriding descriptor was explained in "Overriding Descriptors" on page 898.

❹ Otherwise, the attribute name is unknown, and an exception will be raised by __flag_unknown_attrs.

❺ Build a helpful error message listing all unexpected arguments, and raise Attribu teError. This is a rare example of the NoReturn special type, covered in "NoReturn" on page 295.

❻ Create a dict from the attributes of a Movie object. I'd call this method _as_dict, but I followed the convention started by the _asdict method in col lections.namedtuple.

❼ Implementing a nice __repr__ is the main reason for having _asdict in this example.

The Checked example illustrates how to handle overriding descriptors when implementing __setattr__ to block arbitrary attribute setting after instantiation. It is debatable whether implementing __setattr__ is worthwhile in this example. Without it, setting movie.director = 'Greta Gerwig' would succeed, but the director attribute would not be checked in any way, and would not appear in the __repr__ nor would it be included in the dict returned by _asdict—both defined in Example 24-6.

In *record_factory.py* (Example 24-2) I solved this issue using the __slots__ class attribute. However, this simpler solution is not viable in this case, as explained next.

Why __init_subclass__ Cannot Configure __slots__

The __slots__ attribute is only effective if it is one of the entries in the class namespace passed to type.__new__. Adding __slots__ to an existing class has no effect. Python invokes __init_subclass__ only after the class is built—by then it's too late to configure __slots__. A class decorator can't configure __slots__ either, because it is applied even later than __init_subclass__. We'll explore these timing issues in "What Happens When: Import Time Versus Runtime" on page 929.

To configure __slots__ at runtime, your own code must build the class namespace passed as the last argument of type.__new__. To do that, you can write a class factory function, like *record_factory.py*, or you can take the nuclear option and implement a metaclass. We will see how to dynamically configure __slots__ in "Metaclasses 101" on page 935.

Before PEP 487 (*https://fpy.li/pep487*) simplified the customization of class creation with __init_subclass__ in Python 3.7, similar functionality had to be implemented using a class decorator. That's the focus of the next section.

Enhancing Classes with a Class Decorator

A class decorator is a callable that behaves similarly to a function decorator: it gets the decorated class as an argument, and should return a class to replace the decorated class. Class decorators often return the decorated class itself, after injecting more methods in it via attribute assignment.

Probably the most common reason to choose a class decorator over the simpler __init_subclass__ is to avoid interfering with other class features, such as inheritance and metaclasses.[10]

In this section, we'll study *checkeddeco.py*, which provides the same service as *checkedlib.py*, but using a class decorator. As usual, we'll start by looking at a usage example, extracted from the doctests in *checkeddeco.py* (Example 24-7).

Example 24-7. checkeddeco.py: creating a `Movie` class decorated with `@checked`

```
>>> @checked
... class Movie:
...     title: str
...     year: int
...     box_office: float
...
>>> movie = Movie(title='The Godfather', year=1972, box_office=137)
>>> movie.title
'The Godfather'
>>> movie
Movie(title='The Godfather', year=1972, box_office=137.0)
```

The only difference between Example 24-7 and Example 24-3 is the way the `Movie` class is declared: it is decorated with `@checked` instead of subclassing `Checked`. Otherwise, the external behavior is the same, including the type validation and default value assignments shown after Example 24-3 in "Introducing __init_subclass__" on page 918.

Now let's look at the implementation of *checkeddeco.py*. The imports and `Field` class are the same as in *checkedlib.py*, listed in Example 24-4. There is no other class, only functions in *checkeddeco.py*.

The logic previously implemented in __init_subclass__ is now part of the `checked` function—the class decorator listed in Example 24-8.

---

10 This rationale appears in the abstract of PEP 557–Data Classes (*https://fpy.li/24-9*) to explain why it was implemented as a class decorator.

Example 24-8. checkeddeco.py: the class decorator

```
def checked(cls: type) -> type:     ❶
    for name, constructor in _fields(cls).items():      ❷
        setattr(cls, name, Field(name, constructor))     ❸

    cls._fields = classmethod(_fields)  # type: ignore    ❹

    instance_methods = (     ❺
        __init__,
        __repr__,
        __setattr__,
        _asdict,
        __flag_unknown_attrs,
    )
    for method in instance_methods:     ❻
        setattr(cls, method.__name__, method)

    return cls     ❼
```

❶ Recall that classes are instances of type. These type hints strongly suggest this is a class decorator: it takes a class and returns a class.

❷ _fields is a top-level function defined later in the module (in Example 24-9).

❸ Replacing each attribute returned by _fields with a Field descriptor instance is what __init_subclass__ did in Example 24-5. Here there is more work to do...

❹ Build a class method from _fields, and add it to the decorated class. The type: ignore comment is needed because Mypy complains that type has no _fields attribute.

❺ Module-level functions that will become instance methods of the decorated class.

❻ Add each of the instance_methods to cls.

❼ Return the decorated cls, fulfilling the essential contract of a class decorator.

Every top-level function in *checkeddeco.py* is prefixed with an underscore, except the checked decorator. This naming convention makes sense for a couple of reasons:

- checked is part of the public interface of the *checkeddeco.py* module, but the other functions are not.

- The functions in Example 24-9 will be injected in the decorated class, and the leading _ reduces the chance of naming conflicts with user-defined attributes and methods of the decorated class.

The rest of *checkeddeco.py* is listed in Example 24-9. Those module-level functions have the same code as the corresponding methods of the Checked class of *checkedlib.py*. They were explained in Examples 24-5 and 24-6.

Note that the _fields function does double duty in *checkeddeco.py*. It is used as a regular function in the first line of the checked decorator, and it will also be injected as a class method of the decorated class.

Example 24-9. checkeddeco.py: the methods to be injected in the decorated class

```python
def _fields(cls: type) -> dict[str, type]:
    return get_type_hints(cls)

def __init__(self: Any, **kwargs: Any) -> None:
    for name in self._fields():
        value = kwargs.pop(name, ...)
        setattr(self, name, value)
    if kwargs:
        self.__flag_unknown_attrs(*kwargs)

def __setattr__(self: Any, name: str, value: Any) -> None:
    if name in self._fields():
        cls = self.__class__
        descriptor = getattr(cls, name)
        descriptor.__set__(self, value)
    else:
        self.__flag_unknown_attrs(name)

def __flag_unknown_attrs(self: Any, *names: str) -> NoReturn:
    plural = 's' if len(names) > 1 else ''
    extra = ', '.join(f'{name!r}' for name in names)
    cls_name = repr(self.__class__.__name__)
    raise AttributeError(f'{cls_name} has no attribute{plural} {extra}')

def _asdict(self: Any) -> dict[str, Any]:
    return {
        name: getattr(self, name)
        for name, attr in self.__class__.__dict__.items()
        if isinstance(attr, Field)
    }
```

```
    }

def __repr__(self: Any) -> str:
    kwargs = ', '.join(
        f'{key}={value!r}' for key, value in self._asdict().items()
    )
    return f'{self.__class__.__name__}({kwargs})'
```

The *checkeddeco.py* module implements a simple but usable class decorator. Python's @dataclass does a lot more. It supports many configuration options, adds more methods to the decorated class, handles or warns about conflicts with user-defined methods in the decorated class, and even traverses the __mro__ to collect user-defined attributes declared in the superclasses of the decorated class. The source code (*https://fpy.li/24-10*) of the dataclasses package in Python 3.9 is more than 1,200 lines long.

For metaprogramming classes, we must be aware of when the Python interpreter evaluates each block of code during the construction of a class. This is covered next.

What Happens When: Import Time Versus Runtime

Python programmers talk about "import time" versus "runtime," but the terms are not strictly defined and there is a gray area between them.

At import time, the interpreter:

1. Parses the source code of a *.py* module in one pass from top to bottom. This is when a SyntaxError may occur.

2. Compiles the bytecode to be executed.

3. Executes the top-level code of the compiled module.

If there is an up-to-date *.pyc* file available in the local __pycache__, parsing and compiling are skipped because the bytecode is ready to run.

Although parsing and compiling are definitely "import time" activities, other things may happen at that time, because almost every statement in Python is executable in the sense that they can potentially run user code and may change the state of the user program.

In particular, the import statement is not merely a declaration,[11] but it actually runs all the top-level code of a module when it is imported for the first time in the process. Further imports of the same module will use a cache, and then the only effect will be

---

11 Contrast with the import statement in Java, which is just a declaration to let the compiler know that certain packages are required.

binding the imported objects to names in the client module. That top-level code may do anything, including actions typical of "runtime," such as writing to a log or connecting to a database.[12] That's why the border between "import time" and "runtime" is fuzzy: the import statement can trigger all sorts of "runtime" behavior. Conversely, "import time" can also happen deep inside runtime, because the import statement and the __import__() built-in can be used inside any regular function.

This is all rather abstract and subtle, so let's do some experiments to see what happens when.

Evaluation Time Experiments

Consider an *evaldemo.py* script that uses a class decorator, a descriptor, and a class builder based on __init_subclass__, all defined in a *builderlib.py* module. The modules have several print calls to show what happens under the covers. Otherwise, they don't perform anything useful. The goal of these experiments is to observe the order in which these print calls happen.

 Applying a class decorator and a class builder with __init_sub class__ together in single class is likely a sign of overengineering or desperation. This unusual combination is useful in these experiments to show the timing of the changes that a class decorator and __init_subclass__ can apply to a class.

Let's start by checking out *builderlib.py*, split into two parts: Example 24-10 and Example 24-11.

Example 24-10. builderlib.py: top of the module

```
print('@ builderlib module start')

class Builder:  ❶
    print('@ Builder body')

    def __init_subclass__(cls):  ❷
        print(f'@ Builder.__init_subclass__({cls!r})')

        def inner_0(self):  ❸
            print(f'@ SuperA.__init_subclass__:inner_0({self!r})')

        cls.method_a = inner_0
```

---

12 I'm not saying opening a database connection just because a module is imported is a good idea, only pointing out it can be done.

```
    def __init__(self):
        super().__init__()
        print(f'@ Builder.__init__({self!r})')


def deco(cls):  ❹
    print(f'@ deco({cls!r})')

    def inner_1(self):  ❺
        print(f'@ deco:inner_1({self!r})')

    cls.method_b = inner_1
    return cls  ❻
```

❶ This is a class builder to implement...

❷ ...an __init_subclass__ method.

❸ Define a function to be added to the subclass in the assignment below.

❹ A class decorator.

❺ Function to be added to the decorated class.

❻ Return the class received as an argument.

Continuing with *builderlib.py* in Example 24-11...

Example 24-11. builderlib.py: bottom of the module

```
class Descriptor:  ❶
    print('@ Descriptor body')

    def __init__(self):  ❷
        print(f'@ Descriptor.__init__({self!r})')

    def __set_name__(self, owner, name):  ❸
        args = (self, owner, name)
        print(f'@ Descriptor.__set_name__{args!r}')

    def __set__(self, instance, value):  ❹
        args = (self, instance, value)
        print(f'@ Descriptor.__set__{args!r}')

    def __repr__(self):
        return '<Descriptor instance>'


print('@ builderlib module end')
```

❶   A descriptor class to demonstrate when…

❷   …a descriptor instance is created, and when…

❸   …__set_name__ will be invoked during the owner class construction.

❹   Like the other methods, this __set__ doesn't do anything except display its arguments.

If you import *builderlib.py* in the Python console, this is what you get:

```
>>> import builderlib
@ builderlib module start
@ Builder body
@ Descriptor body
@ builderlib module end
```

Note that the lines printed by *builderlib.py* are prefixed with @.

Now let's turn to *evaldemo.py*, which will trigger special methods in *builderlib.py* (Example 24-12).

Example 24-12. evaldemo.py: script to experiment with builderlib.py

```
#!/usr/bin/env python3

from builderlib import Builder, deco, Descriptor

print('# evaldemo module start')

@deco   ❶
class Klass(Builder):   ❷
    print('# Klass body')

    attr = Descriptor()   ❸

    def __init__(self):
        super().__init__()
        print(f'# Klass.__init__({self!r})')

    def __repr__(self):
        return '<Klass instance>'


def main():   ❹
    obj = Klass()
    obj.method_a()
    obj.method_b()
    obj.attr = 999
```

```
if __name__ == '__main__':
    main()

print('# evaldemo module end')
```

❶ Apply a decorator.

❷ Subclass Builder to trigger its __init_subclass__.

❸ Instantiate the descriptor.

❹ This will only be called if the module is run as the main program.

The print calls in *evaldemo.py* show a # prefix. If you open the console again and import *evaldemo.py*, Example 24-13 is the output.

Example 24-13. Console experiment with evaldemo.py

```
>>> import evaldemo
@ builderlib module start  ❶
@ Builder body
@ Descriptor body
@ builderlib module end
# evaldemo module start
# Klass body  ❷
@ Descriptor.__init__(<Descriptor instance>)  ❸
@ Descriptor.__set_name__(<Descriptor instance>,
      <class 'evaldemo.Klass'>, 'attr')         ❹
@ Builder.__init_subclass__(<class 'evaldemo.Klass'>)  ❺
@ deco(<class 'evaldemo.Klass'>)  ❻
# evaldemo module end
```

❶ The top four lines are the result of from builderlib import… . They will not appear if you didn't close the console after the previous experiment, because *builderlib.py* is already loaded.

❷ This signals that Python started reading the body of Klass. At this point, the class object does not exist yet.

❸ The descriptor instance is created and bound to attr in the namespace that Python will pass to the default class object constructor: type.__new__.

❹ At this point, Python's built-in type.__new__ has created the Klass object and calls __set_name__ on each descriptor instance of descriptor classes that provide that method, passing Klass as the owner argument.

⑤   type.__new__ then calls __init_subclass__ on the superclass of Klass, passing Klass as the single argument.

⑥   When type.__new__ returns the class object, Python applies the decorator. In this example, the class returned by deco is bound to Klass in the module namespace.

The implementation of type.__new__ is written in C. The behavior I just described is documented in the "Creating the class object" (*https://fpy.li/24-11*) section of Python's "Data Model" (*https://fpy.li/dtmodel*) reference.

Note that the main() function of *evaldemo.py* (Example 24-12) was not executed in the console session (Example 24-13), therefore no instance of Klass was created. All the action we saw was triggered by "import time" operations: importing builderlib and defining Klass.

If you run *evaldemo.py* as a script, you will see the same output as Example 24-13 with extra lines right before the end. The extra lines are the result of running main() (Example 24-14).

Example 24-14. Running evaldemo.py as a program

```
$ ./evaldemo.py
[... 9 lines omitted ...]
@ deco(<class '__main__.Klass'>)  ❶
@ Builder.__init__(<Klass instance>)  ❷
# Klass.__init__(<Klass instance>)
@ SuperA.__init_subclass__:inner_0(<Klass instance>)  ❸
@ deco:inner_1(<Klass instance>)  ❹
@ Descriptor.__set__(<Descriptor instance>, <Klass instance>, 999)  ❺
# evaldemo module end
```

❶   The top 10 lines—including this one—are the same as shown in Example 24-13.

❷   Triggered by super().__init__() in Klass.__init__.

❸   Triggered by obj.method_a() in main; method_a was injected by SuperA.__init_subclass__.

❹   Triggered by obj.method_b() in main; method_b was injected by deco.

❺   Triggered by obj.attr = 999 in main.

A base class with __init_subclass__ and a class decorator are powerful tools, but they are limited to working with a class already built by type.__new__ under the covers. In the rare occasions when you need to adjust the arguments passed to `type.__new__`, you need a metaclass. That's the final destination of this chapter—and this book.

Metaclasses 101

> [Metaclasses] are deeper magic than 99% of users should ever worry about. If you wonder whether you need them, you don't (the people who actually need them know with certainty that they need them, and don't need an explanation about why).
>
> —Tim Peters, inventor of the Timsort algorithm and prolific Python contributor[13]

A metaclass is a class factory. In contrast with `record_factory` from Example 24-2, a metaclass is written as a class. In other words, a metaclass is a class whose instances are classes. Figure 24-1 depicts a metaclass using the Mills & Gizmos Notation: a mill producing another mill.

Figure 24-1. A metaclass is a class that builds classes.

Consider the Python object model: classes are objects, therefore each class must be an instance of some other class. By default, Python classes are instances of `type`. In other words, `type` is the metaclass for most built-in and user-defined classes:

```
>>> str.__class__
<class 'type'>
>>> from bulkfood_v5 import LineItem
>>> LineItem.__class__
<class 'type'>
```

---

13 Message to comp.lang.python, subject: "Acrimony in c.l.p." (*https://fpy.li/24-12*). This is another part of the same message from December 23, 2002, quoted in the Preface. The TimBot was inspired that day.

```
>>> type.__class__
<class 'type'>
```

To avoid infinite regress, the class of type is type, as the last line shows.

Note that I am not saying that str or LineItem are subclasses of type. What I am saying is that str and LineItem are instances of type. They all are subclasses of object. Figure 24-2 may help you confront this strange reality.

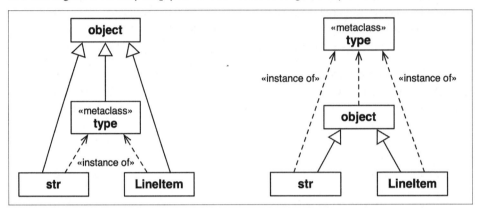

Figure 24-2. Both diagrams are true. The left one emphasizes that str, type, and LineItem are subclasses of object. The right one makes it clear that str, object, and LineItem are instances type, because they are all classes.

The classes object and type have a unique relationship: object is an instance of type, and type is a subclass of object. This relationship is "magic": it cannot be expressed in Python because either class would have to exist before the other could be defined. The fact that type is an instance of itself is also magical.

The next snippet shows that the class of collections.Iterable is abc.ABCMeta. Note that Iterable is an abstract class, but ABCMeta is a concrete class—after all, Iterable is an instance of ABCMeta:

```
>>> from collections.abc import Iterable
>>> Iterable.__class__
<class 'abc.ABCMeta'>
>>> import abc
>>> from abc import ABCMeta
>>> ABCMeta.__class__
<class 'type'>
```

Ultimately, the class of ABCMeta is also type. Every class is an instance of type, directly or indirectly, but only metaclasses are also subclasses of type. That's the most important relationship to understand metaclasses: a metaclass, such as ABCMeta, inherits from type the power to construct classes. Figure 24-3 illustrates this crucial relationship.

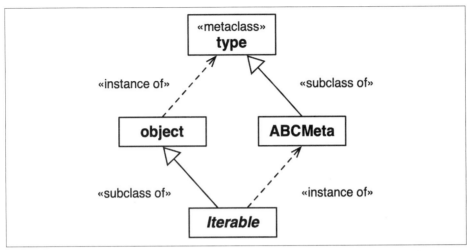

Figure 24-3. Iterable is a subclass of object and an instance of ABCMeta. Both object and ABCMeta are instances of type, but the key relationship here is that ABCMeta is also a subclass of type, because ABCMeta is a metaclass. In this diagram, Iterable is the only abstract class.

The important takeaway here is that metaclasses are subclasses of type, and that's what makes them work as class factories. A metaclass can customize its instances by implementing special methods, as the next sections demonstrate.

How a Metaclass Customizes a Class

To use a metaclass, it's critical to understand how __new__ works on any class. This was discussed in "Flexible Object Creation with __new__" on page 847.

The same mechanics happen at a "meta" level when a metaclass is about to create a new instance, which is a class. Consider this declaration:

```
class Klass(SuperKlass, metaclass=MetaKlass):
    x = 42
    def __init__(self, y):
        self.y = y
```

To process that class statement, Python calls MetaKlass.__new__ with these arguments:

meta_cls
　　The metaclass itself (MetaKlass), because __new__ works as class method.

cls_name

> The string `Klass`.

bases

> The single-element tuple (`SuperKlass,`), with more elements in the case of multiple inheritance.

cls_dict

> A mapping like:

```
{x: 42, `__init__`: <function __init__ at 0x1009c4040>}
```

When you implement `MetaKlass.__new__`, you can inspect and change those arguments before passing them to `super().__new__`, which will eventually call `type.__new__` to create the new class object.

After `super().__new__` returns, you can also apply further processing to the newly created class before returning it to Python. Python then calls `Super Klass.__init_subclass__`, passing the class you created, and then applies a class decorator to it, if one is present. Finally, Python binds the class object to its name in the surrounding namespace—usually the global namespace of a module, if the `class` statement was a top-level statement.

The most common processing made in a metaclass `__new__` is to add or replace items in the `cls_dict`—the mapping that represents the namespace of the class under construction. For instance, before calling `super().__new__`, you can inject methods in the class under construction by adding functions to `cls_dict`. However, note that adding methods can also be done after the class is built, which is why we were able to do it using `__init_subclass__` or a class decorator.

One attribute that you must add to the `cls_dict` before `type.__new__` runs is `__slots__`, as discussed in "Why `__init_subclass__` Cannot Configure `__slots__`" on page 925. The `__new__` method of a metaclass is the ideal place to configure `__slots__`. The next section shows how to do that.

A Nice Metaclass Example

The `MetaBunch` metaclass presented here is a variation of the last example in Chapter 4 of *Python in a Nutshell*, 3rd ed., by Alex Martelli, Anna Ravenscroft, and Steve

Holden, written to run on Python 2.7 and 3.5.[14] Assuming Python 3.6 or later, I was able to further simplify the code.

First, let's see what the Bunch base class provides:

```
>>> class Point(Bunch):
...         x = 0.0
...         y = 0.0
...         color = 'gray'
...
>>> Point(x=1.2, y=3, color='green')
Point(x=1.2, y=3, color='green')
>>> p = Point()
>>> p.x, p.y, p.color
(0.0, 0.0, 'gray')
>>> p
Point()
```

Remember that Checked assigns names to the Field descriptors in subclasses based on class variable type hints, which do not actually become attributes on the class since they don't have values.

Bunch subclasses, on the other hand, use actual class attributes with values, which then become the default values of the instance attributes. The generated __repr__ omits the arguments for attributes that are equal to the defaults.

MetaBunch—the metaclass of Bunch—generates __slots__ for the new class from the class attributes declared in the user's class. This blocks the instantiation and later assignment of undeclared attributes:

```
>>> Point(x=1, y=2, z=3)
Traceback (most recent call last):
    ...
AttributeError: No slots left for: 'z'
>>> p = Point(x=21)
>>> p.y = 42
>>> p
Point(x=21, y=42)
>>> p.flavor = 'banana'
Traceback (most recent call last):
    ...
AttributeError: 'Point' object has no attribute 'flavor'
```

---

14 The authors kindly gave me permission to use their example. MetaBunch first appeared in a message posted by Martelli in the comp.lang.python group on July 7, 2002, with the subject line "a nice metaclass example (was Re: structs in python)" (*https://fpy.li/24-13*), following a discussion about record-like data structures in Python. Martelli's original code for Python 2.2 still runs after a single change: to use a metaclass in Python 3, you must use the metaclass keyword argument in the class declaration, e.g., Bunch(metaclass=MetaBunch), instead of the older convention of adding a __metaclass__ class-level attribute.

Now let's dive into the elegant code of `MetaBunch` in Example 24-15.

Example 24-15. metabunch/from3.6/bunch.py: MetaBunch metaclass and Bunch class

```python
class MetaBunch(type):                                       ❶
    def __new__(meta_cls, cls_name, bases, cls_dict):        ❷

        defaults = {}                                        ❸

        def __init__(self, **kwargs):                        ❹
            for name, default in defaults.items():           ❺
                setattr(self, name, kwargs.pop(name, default))
            if kwargs:                                       ❻
                extra = ', '.join(kwargs)
                raise AttributeError(f'No slots left for: {extra!r}')

        def __repr__(self):                                  ❼
            rep = ', '.join(f'{name}={value!r}'
                            for name, default in defaults.items()
                            if (value := getattr(self, name)) != default)
            return f'{cls_name}({rep})'

        new_dict = dict(__slots__=[], __init__=__init__, __repr__=__repr__)  ❽

        for name, value in cls_dict.items():                 ❾
            if name.startswith('__') and name.endswith('__'):  ❿
                if name in new_dict:
                    raise AttributeError(f"Can't set {name!r} in {cls_name!r}")
                new_dict[name] = value
            else:                                            ⓫
                new_dict['__slots__'].append(name)
                defaults[name] = value
        return super().__new__(meta_cls, cls_name, bases, new_dict)  ⓬


class Bunch(metaclass=MetaBunch):                            ⓭
    pass
```

❶ To create a new metaclass, inherit from `type`.

❷ `__new__` works as a class method, but the class is a metaclass, so I like to name the first argument `meta_cls` (`mcs` is a common alternative). The remaining three arguments are the same as the three-argument signature for calling `type()` directly to create a class.

❸ `defaults` will hold a mapping of attribute names and their default values.

❹ This will be injected into the new class.

❺ Read the `defaults` and set the corresponding instance attribute with a value popped from `kwargs` or a default.

❻ If there is still any item in `kwargs`, it means there are no slots left where we can place them. We believe in *failing fast* as best practice, so we don't want to silently ignore extra items. A quick and effective solution is to pop one item from `kwargs` and try to set it on the instance, triggering an `AttributeError` on purpose.

❼ `__repr__` returns a string that looks like a constructor call—e.g., `Point(x=3)`, omitting the keyword arguments with default values.

❽ Initialize namespace for the new class.

❾ Iterate over the namespace of the user's class.

❿ If a dunder `name` is found, copy the item to the new class namespace, unless it's already there. This prevents users from overwriting `__init__`, `__repr__`, and other attributes set by Python, such as `__qualname__` and `__module__`.

⓫ If not a dunder `name`, append to `__slots__` and save its `value` in `defaults`.

⓬ Build and return the new class.

⓭ Provide a base class, so users don't need to see `MetaBunch`.

`MetaBunch` works because it is able to configure `__slots__` before calling `super().__new__` to build the final class. As usual when metaprogramming, understanding the sequence of actions is key. Let's do another evaluation time experiment, now with a metaclass.

Metaclass Evaluation Time Experiment

This is a variation of "Evaluation Time Experiments" on page 930, adding a metaclass to the mix. The *builderlib.py* module is the same as before, but the main script is now *evaldemo_meta.py*, listed in Example 24-16.

Example 24-16. evaldemo_meta.py: experimenting with a metaclass

```
#!/usr/bin/env python3

from builderlib import Builder, deco, Descriptor
from metalib import MetaKlass     ❶

print('# evaldemo_meta module start')
```

```
@deco
class Klass(Builder, metaclass=MetaKlass):  ❷
    print('# Klass body')

    attr = Descriptor()

    def __init__(self):
        super().__init__()
        print(f'# Klass.__init__({self!r})')

    def __repr__(self):
        return '<Klass instance>'


def main():
    obj = Klass()
    obj.method_a()
    obj.method_b()
    obj.method_c()  ❸
    obj.attr = 999


if __name__ == '__main__':
    main()

print('# evaldemo_meta module end')
```

❶ Import MetaKlass from *metalib.py*, which we'll see in Example 24-18.

❷ Declare Klass as a subclass of Builder and an instance of MetaKlass.

❸ This method is injected by MetaKlass.__new__, as we'll see.

 In the interest of science, Example 24-16 defies all reason and applies three different metaprogramming techniques together on Klass: a decorator, a base class using __init_subclass__, and a custom metaclass. If you do this in production code, please don't blame me. Again, the goal is to observe the order in which the three techniques interfere in the class construction process.

As in the previous evaluation time experiment, this example does nothing but print messages revealing the flow of execution. Example 24-17 shows the code for the top part of *metalib.py*—the rest is in Example 24-18.

Example 24-17. metalib.py: the NosyDict class

```
print('% metalib module start')
```

```
import collections

class NosyDict(collections.UserDict):
    def __setitem__(self, key, value):
        args = (self, key, value)
        print(f'% NosyDict.__setitem__{args!r}')
        super().__setitem__(key, value)

    def __repr__(self):
        return '<NosyDict instance>'
```

I wrote the NosyDict class to override __setitem__ to display each key and value as they are set. The metaclass will use a NosyDict instance to hold the namespace of the class under construction, revealing more of Python's inner workings.

The main attraction of *metalib.py* is the metaclass in Example 24-18. It implements the __prepare__ special method, a class method that Python only invokes on meta-classes. The __prepare__ method provides the earliest opportunity to influence the process of creating a new class.

 When coding a metaclass, I find it useful to adopt this naming convention for special method arguments:

- Use cls instead of self for instance methods, because the instance is a class.

- Use meta_cls instead of cls for class methods, because the class is a metaclass. Recall that __new__ behaves as a class method even without the @classmethod decorator.

Example 24-18. metalib.py: the MetaKlass

```
class MetaKlass(type):
    print('% MetaKlass body')

    @classmethod  ❶
    def __prepare__(meta_cls, cls_name, bases):  ❷
        args = (meta_cls, cls_name, bases)
        print(f'% MetaKlass.__prepare__{args!r}')
        return NosyDict()  ❸

    def __new__(meta_cls, cls_name, bases, cls_dict):  ❹
        args = (meta_cls, cls_name, bases, cls_dict)
        print(f'% MetaKlass.__new__{args!r}')
        def inner_2(self):
            print(f'% MetaKlass.__new__:inner_2({self!r})')

        cls = super().__new__(meta_cls, cls_name, bases, cls_dict.data)  ❺
```

```
        cls.method_c = inner_2  ❻

        return cls  ❼

    def __repr__(cls):  ❽
        cls_name = cls.__name__
        return f"<class {cls_name!r} built by MetaKlass>"

print('% metalib module end')
```

❶ __prepare__ should be declared as a class method. It is not an instance method because the class under construction does not exist yet when Python calls __prepare__.

❷ Python calls __prepare__ on a metaclass to obtain a mapping to hold the namespace of the class under construction.

❸ Return NosyDict instance to be used as the namespace.

❹ cls_dict is a NosyDict instance returned by __prepare__.

❺ type.__new__ requires a real dict as the last argument, so I give it the data attribute of NosyDict, inherited from UserDict.

❻ Inject a method in the newly created class.

❼ As usual, __new__ must return the object just created—in this case, the new class.

❽ Defining __repr__ on a metaclass allows customizing the repr() of class objects.

The main use case for __prepare__ before Python 3.6 was to provide an OrderedDict to hold the attributes of the class under construction, so that the metaclass __new__ could process those attributes in the order in which they appear in the source code of the user's class definition. Now that dict preserves the insertion order, __prepare__ is rarely needed. You will see a creative use for it in "A Metaclass Hack with __prepare__" on page 954.

Importing *metalib.py* in the Python console is not very exciting. Note the use of % to prefix the lines output by this module:

```
>>> import metalib
% metalib module start
% MetaKlass body
% metalib module end
```

Lots of things happen if you import *evaldemo_meta.py*, as you can see in Example 24-19.

Example 24-19. Console experiment with evaldemo_meta.py

```
>>> import evaldemo_meta
@ builderlib module start
@ Builder body
@ Descriptor body
@ builderlib module end
% metalib module start
% MetaKlass body
% metalib module end
# evaldemo_meta module start   ❶
% MetaKlass.__prepare__(<class 'metalib.MetaKlass'>, 'Klass',   ❷
                        (<class 'builderlib.Builder'>,))
% NosyDict.__setitem__(<NosyDict instance>, '__module__', 'evaldemo_meta')   ❸
% NosyDict.__setitem__(<NosyDict instance>, '__qualname__', 'Klass')
# Klass body
@ Descriptor.__init__(<Descriptor instance>)   ❹
% NosyDict.__setitem__(<NosyDict instance>, 'attr', <Descriptor instance>)   ❺
% NosyDict.__setitem__(<NosyDict instance>, '__init__',
                       <function Klass.__init__ at …>)   ❻
% NosyDict.__setitem__(<NosyDict instance>, '__repr__',
                       <function Klass.__repr__ at …>)
% NosyDict.__setitem__(<NosyDict instance>, '__classcell__', <cell at …: empty>)
% MetaKlass.__new__(<class 'metalib.MetaKlass'>, 'Klass',
                    (<class 'builderlib.Builder'>,), <NosyDict instance>)   ❼
@ Descriptor.__set_name__(<Descriptor instance>,
                          <class 'Klass' built by MetaKlass>, 'attr')   ❽
@ Builder.__init_subclass__(<class 'Klass' built by MetaKlass>)
@ deco(<class 'Klass' built by MetaKlass>)
# evaldemo_meta module end
```

❶ The lines before this are the result of importing *builderlib.py* and *metalib.py*.

❷ Python invokes __prepare__ to start processing a class statement.

❸ Before parsing the class body, Python adds the __module__ and __qualname__ entries to the namespace of the class under construction.

❹ The descriptor instance is created…

❺ …and bound to attr in the class namespace.

❻ __init__ and __repr__ methods are defined and added to the namespace.

❼ Once Python finishes processing the class body, it calls MetaKlass.__new__.

❽ __set_name__, __init_subclass__, and the decorator are invoked in this order, after the __new__ method of the metaclass returns the newly constructed class.

If you run *evaldemo_meta.py* as script, main() is called, and a few more things happen (Example 24-20).

Example 24-20. Running evaldemo_meta.py as a program

```
$ ./evaldemo_meta.py
[... 20 lines omitted ...]
@ deco(<class 'Klass' built by MetaKlass>)  ❶
@ Builder.__init__(<Klass instance>)
# Klass.__init__(<Klass instance>)
@ SuperA.__init_subclass__:inner_0(<Klass instance>)
@ deco:inner_1(<Klass instance>)
% MetaKlass.__new__:inner_2(<Klass instance>)  ❷
@ Descriptor.__set__(<Descriptor instance>, <Klass instance>, 999)
# evaldemo_meta module end
```

❶ The top 21 lines—including this one—are the same shown in Example 24-19.

❷ Triggered by obj.method_c() in main; method_c was injected by Meta Klass.__new__.

Let's now go back to the idea of the Checked class with the Field descriptors implementing runtime type validation, and see how it can be done with a metaclass.

A Metaclass Solution for Checked

I don't want to encourage premature optimization and overengineering, so here is a make-believe scenario to justify rewriting *checkedlib.py* with __slots__, which requires the application of a metaclass. Feel free to skip it.

A Bit of Storytelling

Our *checkedlib.py* using __init_subclass__ is a company-wide success, and our production servers have millions of instances of Checked subclasses in memory at any one time.

Profiling a proof-of-concept, we discover that using __slots__ will reduce the cloud hosting bill for two reasons:

- Lower memory usage, as Checked instances don't need their own __dict__

- Higher performance, by removing __setattr__, which was created just to block unexpected attributes, but is triggered at instantiation and for all attribute setting before Field.__set__ is called to do its job

---

The *metaclass/checkedlib.py* module we'll study next is a drop-in replacement for *init-sub/checkedlib.py*. The doctests embedded in them are identical, as well as the *checkedlib_test.py* files for *pytest*.

The complexity in *checkedlib.py* is abstracted away from the user. Here is the source code of a script using the package:

```python
from checkedlib import Checked

class Movie(Checked):
    title: str
    year: int
    box_office: float

if __name__ == '__main__':
    movie = Movie(title='The Godfather', year=1972, box_office=137)
    print(movie)
    print(movie.title)
```

That concise Movie class definition leverages three instances of the Field validating descriptor, a __slots__ configuration, five methods inherited from Checked, and a metaclass to put it all together. The only visible part of checkedlib is the Checked base class.

Consider Figure 24-4. The Mills & Gizmos Notation complements the UML class diagram by making the relationship between classes and instances more visible.

For example, a Movie class using the new *checkedlib.py* is an instance of CheckedMeta, and a subclass of Checked. Also, the title, year, and box_office class attributes of Movie are three separate instances of Field. Each Movie instance has its own _title, _year, and _box_office attributes, to store the values of the corresponding fields.

Now let's study the code, starting with the Field class, shown in Example 24-21.

The Field descriptor class is now a bit different. In the previous examples, each Field descriptor instance stored its value in the managed instance using an attribute of the same name. For example, in the Movie class, the title descriptor stored the field value in a title attribute in the managed instance. This made it unnecessary for Field to provide a __get__ method.

However, when a class like Movie uses __slots__, it cannot have class attributes and instance attributes with the same name. Each descriptor instance is a class attribute, and now we need separate per-instance storage attributes. The code uses the descriptor name prefixed with a single _. Therefore Field instances have separate name and storage_name attributes, and we implement Field.__get__.

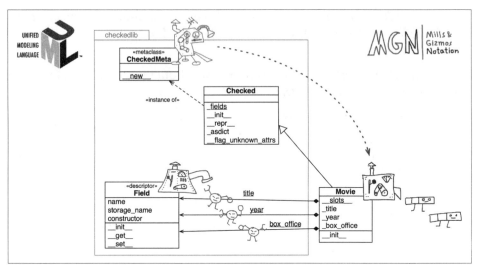

Figure 24-4. UML class diagram annotated with MGN: the `CheckedMeta` meta-mill builds the `Movie` mill. The `Field` mill builds the `title`, `year`, and `box_office` descriptors, which are class attributes of `Movie`. The per-instance data for the fields is stored in the `_title`, `_year`, and `_box_office` instance attributes of `Movie`. Note the package boundary of `checkedlib`. The developer of `Movie` doesn't need to grok all the machinery inside checkedlib.py.

Example 24-21 shows the source code for `Field`, with callouts describing only the changes in this version.

Example 24-21. metaclass/checkedlib.py: the `Field` descriptor with `storage_name` and `__get__`

```python
class Field:
    def __init__(self, name: str, constructor: Callable) -> None:
        if not callable(constructor) or constructor is type(None):
            raise TypeError(f'{name!r} type hint must be callable')
        self.name = name
        self.storage_name = '_' + name        ❶
        self.constructor = constructor

    def __get__(self, instance, owner=None):
        if instance is None:                   ❷
            return self
        return getattr(instance, self.storage_name)   ❸

    def __set__(self, instance: Any, value: Any) -> None:
        if value is ...:
            value = self.constructor()
        else:
```

```
            try:
                value = self.constructor(value)
            except (TypeError, ValueError) as e:
                type_name = self.constructor.__name__
                msg = f'{value!r} is not compatible with {self.name}:{type_name}'
                raise TypeError(msg) from e
        setattr(instance, self.storage_name, value)    ❹
```

❶ Compute storage_name from the name argument.

❷ If __get__ gets None as the instance argument, the descriptor is being read from the managed class itself, not a managed instance. So we return the descriptor.

❸ Otherwise, return the value stored in the attribute named storage_name.

❹ __set__ now uses setattr to set or update the managed attribute.

Example 24-22 shows the code for the metaclass that drives this example.

Example 24-22. metaclass/checkedlib.py: the CheckedMeta metaclass

```
class CheckedMeta(type):

    def __new__(meta_cls, cls_name, bases, cls_dict):    ❶
        if '__slots__' not in cls_dict:    ❷
            slots = []
            type_hints = cls_dict.get('__annotations__', {})    ❸
            for name, constructor in type_hints.items():    ❹
                field = Field(name, constructor)    ❺
                cls_dict[name] = field    ❻
                slots.append(field.storage_name)    ❼

            cls_dict['__slots__'] = slots    ❽

        return super().__new__(
            meta_cls, cls_name, bases, cls_dict)    ❾
```

❶ __new__ is the only method implemented in CheckedMeta.

❷ Only enhance the class if its cls_dict doesn't include __slots__. If __slots__ is already present, assume it is the Checked base class and not a user-defined sub-class, and build the class as is.

❸ To get the type hints in prior examples, we used typing.get_type_hints, but that requires an existing class as the first argument. At this point, the class we are configuring does not exist yet, so we need to retrieve the __annotations__ directly from the `cls_dict`—the namespace of the class under construction, which Python passes as the last argument to the metaclass `__new__`.

❹ Iterate over `type_hints` to…

❺ …build a `Field` for each annotated attribute…

❻ …overwrite the corresponding entry in `cls_dict` with the `Field` instance…

❼ …and append the `storage_name` of the field in the list we'll use to…

❽ …populate the `__slots__` entry in `cls_dict`—the namespace of the class under construction.

❾ Finally, we call `super().__new__`.

The last part of *metaclass/checkedlib.py* is the `Checked` base class that users of this library will subclass to enhance their classes, like `Movie`.

The code for this version of `Checked` is the same as `Checked` in *initsub/checkedlib.py* (listed in Example 24-5 and Example 24-6), with three changes:

1. Added an empty `__slots__` to signal to `CheckedMeta.__new__` that this class doesn't require special processing.

2. Removed `__init_subclass__`. Its job is now done by `CheckedMeta.__new__`.

3. Removed `__setattr__`. It became redundant because adding `__slots__` to the user-defined class prevents setting undeclared attributes.

Example 24-23 is a complete listing of the final version of `Checked`.

Example 24-23. metaclass/checkedlib.py: the `Checked` base class

```python
class Checked(metaclass=CheckedMeta):
    __slots__ = ()  # skip CheckedMeta.__new__ processing

    @classmethod
    def _fields(cls) -> dict[str, type]:
        return get_type_hints(cls)

    def __init__(self, **kwargs: Any) -> None:
        for name in self._fields():
            value = kwargs.pop(name, ...)
            setattr(self, name, value)
        if kwargs:
            self.__flag_unknown_attrs(*kwargs)
```

```
def __flag_unknown_attrs(self, *names: str) -> NoReturn:
    plural = 's' if len(names) > 1 else ''
    extra = ', '.join(f'{name!r}' for name in names)
    cls_name = repr(self.__class__.__name__)
    raise AttributeError(f'{cls_name} object has no attribute{plural} {extra}')

def _asdict(self) -> dict[str, Any]:
    return {
        name: getattr(self, name)
        for name, attr in self.__class__.__dict__.items()
        if isinstance(attr, Field)
    }

def __repr__(self) -> str:
    kwargs = ', '.join(
        f'{key}={value!r}' for key, value in self._asdict().items()
    )
    return f'{self.__class__.__name__}({kwargs})'
```

This concludes the third rendering of a class builder with validated descriptors.

The next section covers some general issues related to metaclasses.

Metaclasses in the Real World

Metaclasses are powerful, but tricky. Before deciding to implement a metaclass, consider the following points.

Modern Features Simplify or Replace Metaclasses

Over time, several common use cases of metaclasses were made redundant by new language features:

Class decorators
> Simpler to understand than metaclasses, and less likely to cause conflicts with base classes and metaclasses.

__set_name__
> Avoids the need for custom metaclass logic to automatically set the name of a descriptor.[15]

---

15 In the first edition of *Fluent Python*, the more advanced versions of the LineItem class used a metaclass just to set the storage name of the attributes. See the code in the metaclasses of bulkfood in the first edition code repository (*https://fpy.li/24-14*).

`__init_subclass__`
> Provides a way to customize class creation that is transparent to the end user and even simpler than a decorator—but may introduce conflicts in a complex class hierarchy.

Built-in `dict` *preserving key insertion order*
> Eliminated the #1 reason to use `__prepare__`: to provide an `OrderedDict` to store the namespace of the class under construction. Python only calls `__prepare__` on metaclasses, so if you needed to process the class namespace in the order it appears in the source code, you had to use a metaclass before Python 3.6.

As of 2021, every actively maintained version of CPython supports all the features just listed.

I keep advocating these features because I see too much unnecessary complexity in our profession, and metaclasses are a gateway to complexity.

Metaclasses Are Stable Language Features

Metaclasses were introduced in Python 2.2 in 2002, together with so-called "new-style classes," descriptors, and properties.

It is remarkable that the `MetaBunch` example, first posted by Alex Martelli in July 2002, still works in Python 3.9—the only change being the way to specify the metaclass to use, which in Python 3 is done with the syntax `class Bunch(metaclass=Meta Bunch):`.

None of the additions I mentioned in "Modern Features Simplify or Replace Metaclasses" on page 951 broke existing code using metaclasses. But legacy code using metaclasses can often be simplified by leveraging those features, especially if you can drop support to Python versions before 3.6—which are no longer maintained.

A Class Can Only Have One Metaclass

If your class declaration involves two or more metaclasses, you will see this puzzling error message:

```
TypeError: metaclass conflict: the metaclass of a derived class
must be a (non-strict) subclass of the metaclasses of all its bases
```

This may happen even without multiple inheritance. For example, a declaration like this could trigger that `TypeError`:

```
class Record(abc.ABC, metaclass=PersistentMeta):
    pass
```

We saw that `abc.ABC` is an instance of the `abc.ABCMeta` metaclass. If that `Persistent` metaclass is not itself a subclass of `abc.ABCMeta`, you get a metaclass conflict.

There are two ways of dealing with that error:

- Find some other way of doing what you need to do, while avoiding at least one of the metaclasses involved.

- Write your own `PersistentABCMeta` metaclass as a subclass of both `abc.ABCMeta` and `PersistentMeta`, using multiple inheritance, and use that as the only metaclass for `Record`.[16]

I can imagine the solution of the metaclass with two base metaclasses implemented to meet a deadline. In my experience, metaclass programming always takes longer than anticipated, which makes this approach risky before a hard deadline. If you do it and make the deadline, the code may contain subtle bugs. Even in the absence of known bugs, you should consider this approach as technical debt simply because it is hard to understand and maintain.

Metaclasses Should Be Implementation Details

Besides `type`, there are only six metaclasses in the entire Python 3.9 standard library. The better known metaclasses are probably `abc.ABCMeta`, `typing.NamedTupleMeta`, and `enum.EnumMeta`. None of them are intended to appear explicitly in user code. We may consider them implementation details.

Although you can do some really wacky metaprogramming with metaclasses, it's best to heed the principle of least astonishment (*https://fpy.li/24-15*) so that most users can indeed regard metaclasses as implementation details.[17]

In recent years, some metaclasses in the Python standard library were replaced by other mechanisms, without breaking the public API of their packages. The simplest way to future-proof such APIs is to offer a regular class that users subclass to access the functionality provided by the metaclass, as we've done in our examples.

To wrap up our coverage of class metaprogramming, I will share with you the coolest, small example of metaclass I found as I researched this chapter.

---

16 If you just got dizzy considering the implications of multiple inheritance with metaclasses, good for you. I'd stay way from this solution as well.

17 I made a living writing Django code for a few years before I decided to study how Django's model fields were implemented. Only then I learned about descriptors and metaclasses.

A Metaclass Hack with __prepare__

When I updated this chapter for the second edition, I needed to find simple but illuminating examples to replace the *bulkfood* LineItem code that no longer require metaclasses since Python 3.6.

The simplest and most interesting metaclass idea was given to me by João S. O. Bueno—better known as JS in the Brazilian Python community. One application of his idea is to create a class that autogenerates numeric constants:

```
>>> class Flavor(AutoConst):
...     banana
...     coconut
...     vanilla
...
>>> Flavor.vanilla
2
>>> Flavor.banana, Flavor.coconut
(0, 1)
```

Yes, that code works as shown! That's actually a doctest in *autoconst_demo.py*.

Here is the user-friendly AutoConst base class and the metaclass behind it, implemented in *autoconst.py*:

```
class AutoConstMeta(type):
    def __prepare__(name, bases, **kwargs):
        return WilyDict()

class AutoConst(metaclass=AutoConstMeta):
    pass
```

That's it.

Clearly the trick is in WilyDict.

When Python processes the namespace of the user's class and reads banana, it looks up that name in the mapping provided by __prepare__: an instance of WilyDict. WilyDict implements __missing__, covered in "The __missing__ Method" on page 91. The WilyDict instance initially has no 'banana' key, so the __missing__ method is triggered. It makes an item on the fly with the key 'banana' and the value 0, returning that value. Python is happy with that, then tries to retrieve 'coconut'. Wily Dict promptly adds that entry with the value 1, returning it. The same happens with 'vanilla', which is then mapped to 2.

We've seen __prepare__ and __missing__ before. The real innovation is how JS put them together.

Here is the source code for WilyDict, also from *autoconst.py*:

```
class WilyDict(dict):
    def __init__(self, *args, **kwargs):
        super().__init__(*args, **kwargs)
        self.__next_value = 0

    def __missing__(self, key):
        if key.startswith('__') and key.endswith('__'):
            raise KeyError(key)
        self[key] = value = self.__next_value
        self.__next_value += 1
        return value
```

While experimenting, I found that Python looked up __name__ in the namespace of the class under construction, causing WilyDict to add a __name__ entry, and increment __next_value. So I added that if statement in __missing__ to raise KeyError for keys that look like dunder attributes.

The *autoconst.py* package both requires and illustrates mastery of Python's dynamic class building machinery.

I had a great time adding more functionality to AutoConstMeta and AutoConst, but instead of sharing my experiments, I will let you have fun playing with JS's ingenious hack.

Here are some ideas:

- Make it possible to retrieve the constant name if you have the value. For example, Flavor[2] could return 'vanilla'. You can to this by implementing __getitem__ in AutoConstMeta. Since Python 3.9, you can implement __class_getitem__ (*https://fpy.li/24-16*) in AutoConst itself.

- Support iteration over the class, by implementing __iter__ on the metaclass. I would make the __iter__ yield the constants as (name, value) pairs.

- Implement a new Enum variant. This would be a major undertaking, because the enum package is full of tricks, including the EnumMeta metaclass with hundreds of lines of code and a nontrivial __prepare__ method.

Enjoy!

The __class_getitem__ special method was added in Python 3.9 to support generic types, as part of PEP 585—Type Hinting Generics In Standard Collections (*https://fpy.li/pep585*). Thanks to __class_getitem__, Python's core developers did not have to write a new metaclass for the built-in types to implement __getitem__ so that we could write generic type hints like list[int]. This is a narrow feature, but representative of a wider use case for metaclasses: implementing operators and other special methods to work at the class level, such as making the class itself iterable, just like Enum subclasses.

Wrapping Up

Metaclasses, as well as class decorators and __init_subclass__ are useful for:

- Subclass registration
- Subclass structural validation
- Applying decorators to many methods at once
- Object serialization
- Object-relational mapping
- Object-based persistence
- Implementing special methods at the class level
- Implementing class features found in other languages, such as traits (*https://fpy.li/24-17*) and aspect-oriented programming (*https://fpy.li/24-18*)

Class metaprogramming can also help with performance issues in some cases, by performing tasks at import time that otherwise would execute repeatedly at runtime.

To wrap up, let's recall Alex Martelli's final advice from his essay "Waterfowl and ABCs" on page 445:

> And, *don't* define custom ABCs (or metaclasses) in production code... if you feel the urge to do so, I'd bet it's likely to be a case of "all problems look like a nail"-syndrome for somebody who just got a shiny new hammer—you (and future maintainers of your code) will be much happier sticking with straightforward and simple code, eschewing such depths.

I believe Martelli's advice applies not only to ABCs and metaclasses, but also to class hierarchies, operator overloading, function decorators, descriptors, class decorators, and class builders using __init_subclass__.

Those powerful tools exist primarily to support library and framework development. Applications naturally should *use* those tools, as provided by the Python standard library or external packages. But *implementing* them in application code is often premature abstraction.

> Good frameworks are extracted, not invented.[18]
>
> —David Heinemeier Hansson, creator of Ruby on Rails

Chapter Summary

This chapter started with an overview of attributes found in class objects, such as __qualname__ and the __subclasses__() method. Next, we saw how the type built-in can be used to construct classes at runtime.

The __init_subclass__ special method was introduced, with the first iteration of a Checked base class designed to replace attribute type hints in user-defined subclasses with Field instances that apply constructors to enforce the type of those attributes at runtime.

The same idea was implemented with a @checked class decorator that adds features to user-defined classes, similar to what __init_subclass__ allows. We saw that neither __init_subclass__ nor a class decorator can dynamically configure __slots__, because they operate only after a class is created.

The concepts of "import time" and "runtime" were clarified with experiments showing the order in which Python code is executed when modules, descriptors, class decorators, and __init_subclass__ is involved.

Our coverage of metaclasses began with an overall explanation of type as a metaclass, and how user-defined metaclasses can implement __new__ to customize the classes it builds. We then saw our first custom metaclass, the classic MetaBunch example using __slots__. Next, another evaluation time experiment demonstrated how the __prepare__ and __new__ methods of a metaclass are invoked earlier than __init_subclass__ and class decorators, providing opportunities for deeper class customization.

The third iteration of a Checked class builder with Field descriptors and custom __slots__ configuration was presented, followed by some general considerations about metaclass usage in practice.

Finally, we saw the AutoConst hack invented by João S. O. Bueno, based on the cunning idea of a metaclass with __prepare__ returning a mapping that implements __missing__. In less than 20 lines of code, *autoconst.py* showcases the power of combining Python metaprogramming techniques

---

18 The phrase is widely quoted. I found an early direct quote in a post (*https://fpy.li/24-19*) in DHH's blog from 2005.

I haven't yet found a language that manages to be easy for beginners, practical for professionals, and exciting for hackers in the way that Python is. Thanks, Guido van Rossum and everybody else who makes it so.

Further Reading

Caleb Hattingh—a technical reviewer of this book—wrote the *autoslot* (*https://fpy.li/ 24-20*) package, providing a metaclass to automatically create a __slots__ attribute in a user-defined class by inspecting the bytecode of __init__ and finding all assignments to attributes of self. It's useful and also an excellent example to study: only 74 lines of code in *autoslot.py*, including 20 lines of comments explaining the most difficult parts.

The essential references for this chapter in the Python documentation are "3.3.3. Customizing class creation" (*https://fpy.li/24-21*) in the "Data Model" chapter of *The Python Language Reference*, which covers __init_subclass__ and metaclasses. The type class documentation (*https://fpy.li/24-22*) in the "Built-in Functions" page, and "4.13. Special Attributes" (*https://fpy.li/24-1*) of the "Built-in Types" chapter in the *The Python Standard Library* are also essential reading.

In the *The Python Standard Library*, the types module documentation (*https://fpy.li/ 24-24*) covers two functions added in Python 3.3 that simplify class metaprogramming: types.new_class and types.prepare_class.

Class decorators were formalized in PEP 3129—Class Decorators (*https://fpy.li/ 24-25*), written by Collin Winter, with the reference implementation authored by Jack Diederich. The PyCon 2009 talk "Class Decorators: Radically Simple" (video (*https://fpy.li/24-26*)), also by Jack Diederich, is a quick introduction to the feature. Besides @dataclass, an interesting—and much simpler—example of a class decorator in Python's standard library is functools.total_ordering (*https://fpy.li/24-27*) that generates special methods for object comparison.

For metaclasses, the main reference in Python's documentation is PEP 3115—Metaclasses in Python 3000 (*https://fpy.li/pep3115*), in which the __prepare__ special method was introduced.

Python in a Nutshell, 3rd ed., by Alex Martelli, Anna Ravenscroft, and Steve Holden, is authoritative, but was written before PEP 487—Simpler customization of class creation (*https://fpy.li/pep487*) came out. The main metaclass example in that book—MetaBunch—is still valid, because it can't be written with simpler mechanisms. Brett Slatkin's *Effective Python*, 2nd ed. (Addison-Wesley) has several up-to-date examples of class building techniques, including metaclasses.

To learn about the origins of class metaprogramming in Python, I recommend Guido van Rossum's paper from 2003, "Unifying types and classes in Python 2.2" (*https://fpy.li/24-28*). The text applies to modern Python as well, as it covers what were then called the "new-style" class semantics—the default semantics in Python 3—including descriptors and metaclasses. One of the references cited by Guido is *Putting Metaclasses to Work: a New Dimension in Object-Oriented Programming*, by Ira R. Forman and Scott H. Danforth (Addison-Wesley), a book to which he gave five stars on *Amazon.com*, adding the following review:

> **This book contributed to the design for metaclasses in Python 2.2**
>
> Too bad this is out of print; I keep referring to it as the best tutorial I know for the difficult subject of cooperative multiple inheritance, supported by Python via the super() function.[19]

If you are keen on metaprogramming, you may wish Python had the ultimate metaprogramming feature: syntactic macros, as offered in the Lisp family of languages and —more recently—by Elixir and Rust. Syntactic macros are more powerful and less error prone than the primitive code substitution macros in the C language. They are special functions that rewrite source code using custom syntax into standard code before the compilation step, enabling developers to introduce new language constructs without changing the compiler. Like operator overloading, syntactic macros can be abused. But as long as the community understands and manages the downsides, they support powerful and user-friendly abstractions, like DSLs (Domain-Specific Languages). In September 2020, Python core developer Mark Shannon posted PEP 638—Syntactic Macros (*https://fpy.li/pep638*), advocating just that. A year after it was initially published, PEP 638 was still in draft and there were no ongoing discussions about it. Clearly it's not a top priority for the Python core developers. I would like to see PEP 638 further discussed and eventually approved. Syntactic macros would allow the Python community to experiment with controversial new features, such as the walrus operator (PEP 572 (*https://fpy.li/pep572*)), pattern matching (PEP 634 (*https://fpy.li/pep634*)), and alternative rules for evaluating type hints (PEPs 563 (*https://fpy.li/pep563*) and 649 (*https://fpy.li/pep649*)) before making permanent changes to the core language. Meanwhile, you can get a taste of syntactic macros with the MacroPy (*https://fpy.li/24-29*) package.

---

19 I bought a used copy and found it a very challenging read.

Soapbox

I will start the last soapbox in the book with a long quote from Brian Harvey and Matthew Wright, two computer science professors from the University of California (Berkeley and Santa Barbara). In their book, *Simply Scheme: Introducing Computer Science* (MIT Press), Harvey and Wright wrote:

> There are two schools of thought about teaching computer science. We might caricature the two views this way:
>
> 1. **The conservative view**: Computer programs have become too large and complex to encompass in a human mind. Therefore, the job of computer science education is to teach people how to discipline their work in such a way that 500 mediocre programmers can join together and produce a program that correctly meets its specification.
>
> 2. **The radical view**: Computer programs have become too large and complex to encompass in a human mind. Therefore, the job of computer science education is to teach people how to expand their minds so that the programs can fit, by learning to think in a vocabulary of larger, more powerful, more flexible ideas than the obvious ones. Each unit of programming thought must have a big payoff in the capabilities of the program.
>
> —Brian Harvey and Matthew Wright, preface to *Simply Scheme*[20]

Harvey and Wright's exaggerated descriptions are about teaching computer science, but they also apply to programming language design. By now, you should have guessed that I subscribe to the "radical" view, and I believe Python was designed in that spirit.

The property idea is a great step forward compared to the accessors-from-the-start approach practically demanded by Java and supported by Java IDEs generating getters/setters with a keyboard shortcut. The main advantage of properties is to let us start our programs simply exposing attributes as public—in the spirit of *KISS*—knowing a public attribute can become a property at any time without much pain. But the descriptor idea goes way beyond that, providing a framework for abstracting away repetitive accessor logic. That framework is so effective that essential Python constructs use it behind the scenes.

Another powerful idea is functions as first-class objects, paving the way to higher-order functions. Turns out the combination of descriptors and higher-order functions enable the unification of functions and methods. A function's __get__ produces

---

20 See p. xvii. Full text available at Berkeley.edu (*https://fpy.li/24-30*).

a method object on the fly by binding the instance to the self argument. This is elegant.[21]

Finally, we have the idea of classes as first-class objects. It's an outstanding feat of design that a beginner-friendly language provides powerful abstractions such as class builders, class decorators, and full-fledged, user-defined metaclasses. Best of all, the advanced features are integrated in a way that does not complicate Python's suitability for casual programming (they actually help it, under the covers). The convenience and success of frameworks such as Django and SQLAlchemy owe much to metaclasses. Over the years, class metaprogramming in Python is becoming simpler and simpler, at least for common use cases. The best language features are those that benefit everyone, even if some Python users are not aware of them. But they can always learn and create the next great library.

I look forward to learning about your contributions to the Python community and ecosystem!

---

21 *Machine Beauty: Elegance and the Heart of Technology* by David Gelernter (Basic Books) opens with an intriguing discussion of elegance and aesthetics in works of engineering, from bridges to software. The later chapters are not great, but the opening is worth the price.

Afterword

Python is a language for consenting adults.

—Alan Runyan, cofounder of Plone

Alan's pithy definition expresses one of the best qualities of Python: it gets out of the way and lets you do what you must. This also means it doesn't give you tools to restrict what others can do with your code and the objects it builds.

At age 30, Python is still growing in popularity. But of course, it is not perfect. Among the top irritants to me is the inconsistent use of CamelCase, snake_case, and joinedwords in the standard library. But the language definition and the standard library are only part of an ecosystem. The community of users and contributors is the best part of the Python ecosystem.

Here is one example of the community at its best: while writing about *asyncio* in the first edition, I was frustrated because the API has many functions, dozens of which are coroutines, and you had to call the coroutines with yield from—now with await —but you can't do that with regular functions. This was documented in the *asyncio* pages, but sometimes you had to read a few paragraphs to find out whether a particular function was a coroutine. So I sent a message to python-tulip titled "Proposal: make coroutines stand out in the *asyncio* docs" (*https://fpy.li/a-1*). Victor Stinner, an *asyncio* core developer; Andrew Svetlov, main author of *aiohttp*; Ben Darnell, lead developer of Tornado; and Glyph Lefkowitz, inventor of *Twisted*, joined the conversation. Darnell suggested a solution, Alexander Shorin explained how to implement it in Sphinx, and Stinner added the necessary configuration and markup. Less than 12 hours after I raised the issue, the entire *asyncio* documentation set online was updated with the *coroutine* tags (*https://fpy.li/a-2*) you can see today.

That story did not happen in an exclusive club. Anybody can join the python-tulip list, and I had posted only a few times when I wrote the proposal. The story illustrates a community that is really open to new ideas and new members. Guido van Rossum used to hang out in python-tulip and often answered basic questions.

Another example of openness: the Python Software Foundation (PSF) has been working to increase diversity in the Python community. Some encouraging results are already in. The 2013–2014 PSF board saw the first women elected directors: Jessica McKellar and Lynn Root. In 2015, Diana Clarke chaired PyCon North America in Montréal, where about one-third of the speakers were women. PyLadies became a truly global movement, and I am proud that we have so many PyLadies chapters in Brazil.

If you are a Pythonista but you have not engaged with the community, I encourage you to do so. Seek the PyLadies or Python Users Group (PUG) in your area. If there isn't one, create it. Python is everywhere, so you will not be alone. Travel to events if you can. Join live events too. During the Covid-19 pandemic I learned a lot in the "hallway tracks" of online conferences. Come to a PythonBrasil conference—we've had international speakers regularly for many years now. Hanging out with fellow Pythonistas brings real benefits besides all the knowledge sharing. Like real jobs and real friendships.

I know I could not have written this book without the help of many friends I made over the years in the Python community.

My father, Jairo Ramalho, used to say "Só erra quem trabalha," Portuguese for "Only those who work make mistakes," great advice to avoid being paralyzed by the fear of making errors. I certainly made my share of mistakes while writing this book. The reviewers, editors, and early release readers caught many of them. Within hours of the first edition early release, a reader was reporting typos in the errata page for the book. Other readers contributed more reports, and friends contacted me directly to offer suggestions and corrections. The O'Reilly copyeditors will catch other errors during the production process, which will start as soon as I manage to stop writing. I take responsibility and apologize for any errors and suboptimal prose that remains.

I am very happy to bring this second edition to conclusion, mistakes and all, and I am very grateful to everybody who helped along the way.

I hope to see you soon at some live event. Please come say hi if you see me around!

Further Reading

I will wrap up the book with references regarding what it its to be "Pythonic"—the main question this book tried to address.

Brandon Rhodes is an awesome Python teacher, and his talk "A Python Æsthetic: Beauty and Why I Python" (*https://fpy.li/a-3*) is beautiful, starting with the use of Unicode U+00C6 (LATIN CAPITAL LETTER AE) in the title. Another awesome teacher, Raymond Hettinger, spoke of beauty in Python at PyCon US 2013: "Transforming Code into Beautiful, Idiomatic Python" (*https://fpy.li/a-4*).

The "Evolution of Style Guides" thread (*https://fpy.li/a-5*) that Ian Lee started on Python-ideas is worth reading. Lee is the maintainer of the `pep8` (*https://fpy.li/a-6*) package that checks Python source code for PEP 8 compliance. To check the code in this book, I used `flake8` (*https://fpy.li/a-7*), which wraps `pep8`, `pyflakes` (*https://fpy.li/a-8*), and Ned Batchelder's McCabe complexity plug-in (*https://fpy.li/a-9*).

Besides PEP 8, other influential style guides are the *Google Python Style Guide* (*https://fpy.li/a-10*) and the *Pocoo Styleguide* (*https://fpy.li/a-11*), from the team that brought us Flake, Sphinx, Jinja 2, and other great Python libraries.

The Hitchhiker's Guide to Python! (*https://fpy.li/a-12*) is a collective work about writing Pythonic code. Its most prolific contributor is Kenneth Reitz, a community hero thanks to his beautifully Pythonic `requests` package. David Goodger presented a tutorial at PyCon US 2008 titled "Code Like a Pythonista: Idiomatic Python" (*https://fpy.li/a-13*). If printed, the tutorial notes are 30 pages long. Goodger created both reStructuredText and `docutils`—the foundations of Sphinx, Python's excellent documentation system (which, by the way, is also the official documentation system (*https://fpy.li/a-14*) for MongoDB and many other projects).

Martijn Faassen tackles the question head-on in "What is Pythonic?" (*https://fpy.li/a-15*) In the python-list, there is a thread with that same title (*https://fpy.li/a-16*). Martijn's post is from 2005, and the thread from 2003, but the Pythonic ideal hasn't changed much—neither has the language, for that matter. A great thread with "Pythonic" in the title is "Pythonic way to sum n-th list element?" (*https://fpy.li/a-17*), from which I quoted extensively in the "Soapbox" on page 429.

PEP 3099 — Things that will Not Change in Python 3000 (*https://fpy.li/pep3099*) explains why many things are the way they are, even after the major overhaul that was Python 3. For a long time, Python 3 was nicknamed Python 3000, but it arrived a few centuries sooner—to the dismay of some. PEP 3099 was written by Georg Brandl, compiling many opinions expressed by the *BDFL*, Guido van Rossum. The "Python Essays" (*https://fpy.li/a-18*) page lists several texts by Guido himself.

Index

Symbols

!= (not equal to) operator, 581
!r conversion field, 12
% (modulo) operator, 5, 12
%r placeholder, 12
* (star) operator, 10, 36-37, 50-56, 240, 576-578
** (double star) operator, 80, 240
*= (star equals) operator, 53-56, 584-589
*_ symbol, 42
*_new*_, 847-849
+ operator, 10, 50-56, 570-576
+= (addition assignment) operator, 53-56, 584-589
+ELLIPSIS directive, 7
:= (Walrus operator), 26
< (less than) operator, 581
<= (less than or equal to) operator, 581
== (equality) operator, 206, 225, 581
> (greater than) operator, 581
>= (greater than or equal to) operator, 581
@ sign, 578-580
@asyncio.coroutine decorator, 781
@cached_property, 860
@contextmanager decorator, 668-673
@dataclass
    default settings, 180
    example using, 187-189
    field options, 180-183
    init-only variables, 186
    keyword parameters accepted by, 179
    post-init processing, 183-185
    typed class attributes, 185
    __hash__ method, 180
@typing.overload decorator, 524-530

[] (square brackets), 6, 26, 35, 49
\ (backslash), 26
\ line continuation escape, 26
\N{} (Unicode literals escape notation), 136
_ symbol, 41
__ (double underscore), 3
__abs__, 11
__add__, 11
__bool__, 13
__bytes__, 367
__call__, 362
__class__, 874, 912
__contains__, 7, 93
__delattr__, 876
__delete__, 883
__del__, 219
__dict__, 874
__dir__, 876
__enter__, 662
__eq__, 413-418
__exit__, 662
__format__, 367, 372, 420-427, 430
__getattribute__, 876
__getattr__, 409-413, 876
__getitem__, 5-8, 49, 405-409
__get__, 883
__hash__, 180, 413-418
__iadd__, 53
__init_subclass__, 918-925
__init__, 9, 183, 401
__invert__, 567
__iter__, 607-614
__len__, 5-8, 17, 405-409
__missing__, 91-95

properties (see computed properties; dynamic attributes and properties)
property class, 864-869
protocol classes, 405
Protocol type, 287-292
protocols (see also interfaces)
    defensive programming, 442-444
    duck typing and, 404
    further reading on, 484
    implementing at runtime, 440-442
    implementing generic static protocols, 556-558
    as informal interfaces, 429
    meanings of protocol, 436
    numeric, 480-483
    overview of, 483
    sequence and iterable protocols, 438-440
    significant changes to, 435
    Soapbox discussion, 486-488
    static protocols, 405, 468-483
    topics covered, 435
PSF (see Python Software Foundation)
PUG (see Python Users Group)
PyICU, 150
PyLadies, 964
pytest package, xxii
Python
    appreciating language-specific features, xix
    approach to learning, xx-xxii
    community support for, 963
    fluentpython.com, xxiii
    functional programming with, 250
    functioning with multicore processors, 729-737
    further reading on, 964
    prerequisites to learning, xx
    target audience, xx
    versions featured, xx
Python Data Model
    further reading on, 18
    __getitem__ and __len__, 5-8
    making len work with custom objects, 17
    overview of, 3, 18
    significant changes to, 4
    Soapbox discussion, 19
    special methods overview, 15-17
    using special methods, 8-15
Python Software Foundation (PSF), 964
Python type checkers, 256

Python Users Group (PUG), 964
python-tulip list, 963
Pythonic Card Deck example, 5-8
Pythonic objects (see also objects)
    alternative constructor for, 370
    building user-defined classes, 365
    classmethod versus staticmethod, 371
    formatted displays, 372-376
    further reading on, 394
    hashable Vector2d, 376-379
    object representations, 366
    overriding class attributes, 391-393
    overview of, 393
    private and protected attributes, 384-386
    saving memory with __slots__, 386-390
    significant changes to, 366
    Soapbox discussion, 396-398
    supporting positional patterns, 379
    topics covered, 365
    Vector2d class example, 367-370
    Vector2d full listing, 380-383
Pythonic sums, 430-432
pyuca library, 150, 158

Q

quantity properties, 869-872
questions and comments, xxv
queues
    definition of term, 702
    deque (double-ended queue), 59, 67
    distributed task queues, 736
    implementing, 69

R

race conditions, 727
random.choice function, 6
recycling (see garbage collection)
reduce function, 234-236
reducing functions, 414, 634-635
refactoring strategies
    choosing the best, 352
    classic, 344-348
    Command pattern, 357-359
    decorator-enhanced pattern, 355-357
    finding strategies in modules, 353-355
    function-oriented, 349-352
reference counting, 219
registration decorators, 310, 331-334
regular expressions, str versus bytes in, 155

repr() function, 366
reserved keywords, 683
rich comparison operators, 581-584
running averages, computing, 647-649

S

S-expression, 673
salts, 85
Scheme language, 43-47, 673-675
SciPy, 64-67
scope
    dynamic scope versus lexical scope, 340-341
    function local scope, 312
    module global scope, 312
    variable scope rules, 310-312
    within comprehensions and generator
        expressions, 26
semaphores, 795-799
Sentence classes, 607-614
sentinels, 725
sequence protocol, 438-440, 598-600
sequences
    alternatives to lists, 59-70
    further reading on, 71
    list comprehensions and generator expres-
        sions, 25-30
    list.sort versus sorted built-in, 56-58
    overview of, 70
    overview of built-in, 22-24
    pattern matching with, 39-47
    significant changes to, 22
    slicing, 47-50
    Soapbox discussion, 73-75
    topics covered, 22
    tuples, 30-35
    uniform handling of, 21
    unpacking sequences and iterables, 35-38
    using + and * with, 50-56
sequences, special methods for
    applications beyond three dimensions, 400
    dynamic attribute access, 409-413
    __format__, 420-427
    further reading on, 428
    __hash__ and __eq__, 413-418
    overview of, 427
    protocols and duck typing, 404
    significant changes to, 400
    sliceable sequences, 405-409
    Soapbox discussion, 429-432

topics covered, 399
    Vector implementation strategy, 400
    Vector2d compatibility, 401-403
sequential.py program, 720
server-side web/mobile development, 732
servers
    Asynchronous Server Gateway Interface
        (ASGI), 736
    HTTPServer class, 505
    TCP servers, 808-815
    test servers, 769
    ThreadingHTTPServer class, 505
    Web Server Gateway Interface (WSGI), 734
    writing asyncio servers, 803-815
setattr function, 875
sets (see also dictionaries and sets)
    consequences of how set works, 107
    set comprehensions, 106
    set literals, 105
    set operations, 107-110
    set operations on dict views, 110
    set theory, 103-105
shallow copies, 208-211
shelve module, 97
simple class patterns, 192
single dispatch generic functions, 326-331
Sized interface, 15
slicing
    assigning to slices, 50
    excluding last item in, 47
    multidimensional slicing and ellipses, 49
    slice objects, 48
    sliceable sequences, 405-409
Soapbox sidebars
    @dataclass, 197
    anonymous functions, 252
    __call__, 362
    code points, 161
    data model versus object model, 19
    design patterns, 361
    duck typing, 303, 429
    dynamic scope versus lexical scope, 340-341
    equality (==) operator, 225
    explicit self argument, 909
    flat versus container sequences, 73
    __format__, 430
    function-class duality, 880
    functional programming with Python, 250
    generic collections, 303

representation using special methods, 12
strong testing, 296
struct module, 118
structural typing, 466-468
structured concurrency, 827
subclassing (see inheritance and subclassing)
subgenerators, 636
subtype-of relationships, 267
super() function, 413, 490-492
syntactic sugar, 114
SyntaxError, 128
system administration, 730

T

tail call optimization (TCO), 695-697
TCP servers, 808-815
TensorFlow, 731
test servers, 769
text files, handling, 131-139 (see also Unicode
  text versus bytes)
ThreadingHTTPServer class, 505
ThreadingMixIn class, 505
threads
    definition of term, 702
    enhancing asyncio downloader, 793-801
    further reading on, 738
    Global Interpreter Lock impact, 718
    spinners (loading indicators) using, 705-708
    thread avoidance, 777
    thread-based process pools, 728
throttling, 795-799
Timsort algorithm, 74
Tkinter GUI toolkit
    benefits and drawbacks of, 515
    multiple inheritance in, 509-511
tree structures, traversing, 638-643
tuples
    classic named tuples, 169-172
    immutability and, 221
    as immutable lists, 32-34
    versus lists, 34
    nature of, 73
    as records, 30-32
    relative immutability of, 207
    simplified memory diagram for, 23
    tuple unpacking, 32
    type hints (type annotations), 274-276
    typing.NamedTuple, 172
Twisted library, 835

type casting, 538-541
type hints (type annotations)
    annotating positional only and variadic
      parameters, 295
    for asynchronous objects, 828
    basics of, 173-179
    benefits and drawbacks of, 253
    flawed typing and strong testing, 296
    further reading on, 299
    generic type hints for coroutines, 654
    gradual typing, 254-260 (see also gradual
      type system)
    overview of, 298
    significant changes to, 254
    Soapbox discussion, 300-303
    supported operations and, 261-266
    topics covered, 254
    types usable in, 266-295
typed double function, 468-470
TypedDict, 164, 530-538
Typeshed project, 281
TypeVar, 282-286
typing map, 434, 487 (see also type hints (type
    annotations))
typing module, 266
typing.NamedTuple, 172

U

UCA (see Unicode Collation Algorithm)
UCS-2 encoding, 124
UML class diagrams
    ABCs in collections.abc, 451
    annotated with MGN, 886, 947
    Command design pattern, 357
    django.views.generic.base module, 506
    django.views.generic.list module, 508
    fundamental collection types, 14
    managed and descriptor classes, 884
    MutableSequence ABC and superclasses,
      450
    Sequence ABC and abstract classes, 438
    simplified for collections.abc, 24
    simplified for MutableMapping and super-
      classes, 83
    simplified for MutableSet and superclasses,
      107
    Strategy design pattern, 344
    Tkinter Text widget class and superclasses,
      501

About the Author

Luciano Ramalho was a web developer before the Netscape IPO in 1995, and switched from Perl to Java to Python in 1998. He joined Thoughtworks in 2015, where he is a Principal Consultant in the São Paulo office. He has delivered keynotes, talks, and tutorials at Python events in the Americas, Europe, and Asia, and also presented at Go and Elixir conferences, focusing on language design topics. Ramalho is a fellow of the Python Software Foundation and cofounder of Garoa Hacker Clube, the first hackerspace in Brazil.

Colophon

The animal on the cover of *Fluent Python* is a Namaqua sand lizard (*Pedioplanis namaquensis*), found throughout Namibia in arid savannah and semi-desert regions.

The Namaqua sand lizard has a black body with four white stripes running down its back, brown legs with white spots, a white belly, and a long, pinkish-brown tail. It is one of the fastest of the lizards active during the day and feeds on small insects. It inhabits sparsely vegetated sand gravel flats. Female Namaqua sand lizards lay between three to five eggs in November, and these lizards spends the rest of winter dormant in burrows that they dig near the base of bushes.

The current conservation status of the Namaqua sand lizard is of "Least Concern." Many of the animals on O'Reilly covers are endangered; all of them are important to the world.

The cover illustration is by Karen Montgomery, based on a black and white engraving from Wood's *Natural History*. The cover fonts are Gilroy Semibold and Guardian Sans. The text font is Adobe Minion Pro; the heading font is Adobe Myriad Condensed; and the code font is Dalton Maag's Ubuntu Mono.